WASTEWATER ENGINEERING

WASTEWATER ENGINEERING
TREATMENT AND REUSE

Vikram Bhatt
Mila Kunis

Kruger Brentt
Publishers
2023

Kruger Brentt Publishers UK. LTD.
Company Number 9728962

Regd. Office: 68 St Margarets Road, Edgware, Middlesex HA8 9UU

ISBN: 9781787150843

For information on all our publications visit our website at http://krugerbrentt.com/

PREFACE

Water treatment facilities are designed to speed up the natural process of purifying water. With billions of people and even more wastewater, the natural process is overloaded. Without wastewater treatment, the amount of wastewater would cause devastation, as it still does today in developing countries. Globally, over 80 percent of all wastewater is discharged without treatment.1 In the countries that do have water treatment facilities, they use various methods to treat water with one common goal: purify water as much as possible and send it back into the environment to keep humans and the Earth safe and thriving.Water resources engineering is the study and management of equipment, facilities and techniques that are used to manage and preserve life's most plentiful resource. In addition to assessing how and the best ways in which to control water as it pertains to water-related activities – such as irrigation, waste disposal and canal development – water resource engineers are also frequently involved in water management to ensure that it's safe to drink both for humans, plants and animal usage.

Water resource engineers may be charged with developing new systems or processes for private or government entities that can preserve freshwater sources and find new ones. This may require the assistance of civil engineers involved as well, designing water purification methods through desalination or creating new equipment for contaminant transport when water is used for irrigation purposes. Understanding what works and what doesn't when it comes to water resource management is often a combined effort and may involve a number of different analyses, including hydrologic, which is the study of the water cycle and directions in which it flows, which may be influenced by weather and other environmental forces.

The present book contains nine chapters covering all related disciplines. These chapters include introduction to wastewater engineering,wastewater treatment,structure, properties and uses of water,structure of aquatic system, water conservation methods,water treatment methods,treatment plants, rain water harvesting and management of waste water. It covers the implementation of advanced wastewater treatment techniques, with a focus on pollutant removal. It explores the physical, chemical, and biological treatment technologies

Discusses the advancements of materials and technologies applicable to both potable water and wastewater from industrial municipal sources. This book would be an invaluable asset for the students of chemical, civil and environmental engineering and environmental engineering and also be useful to the practising professionals involved in the design of wastewater treatment plants.

We are grateful to all those persons as well as various books, manuals, periodicals, magazines, journals etc. that helped in the preparation of this book. In spite of the best efforts, it is possible that some errors may have occurred into the compilation and editing of the book. Further queries, constructive suggestions and criticisms for the improvement of the book are always welcome and shall be thankfully acknowledged.

Vikram Bhatt

Mila Kunis

CONTENTS

1

INTRODUCTION TO WASTEWATER ENGINEERING

1.1 OBJECTIVES

After studying this chapter you will be able to understand:

- History of water supply and sanilation.
- Wastewater reuse activities.
- Bronze and early iron ages.
- Ancient mediterranean
- Classic and early modern mesoamerica.
- Application on agricultural land.

1.2 INTRODUCTION

Wastewater engineering, also commonly referred to as sanitary engineering, involves the use of mechanical, biological, and chemical operations to achieve and maintain excellent water quality on a daily basis. Wastewater engineers play a pivotal role in our society, for their rigorous testing allows us to always have pure and clean water to drink and use within our homes and businesses. While there's a specific concentration in the transportation and cleaning of irrigation water, blackwater, and greywater, the role of a wastewater engineer is quite diverse.

Since there's only a limited number of schools that offer degrees specifically in wastewater engineering, many wastewater engineers will receive degrees in civil engineering, environmental engineering, chemical engineering, and even bio-chemical engineering. All high-authority wastewater engineering companies will only consider those who've received their bachelor's degree in the respective engineering program. Many times, a master's degree is required for those who want to attain a higher position–i.e. a wastewater investigator. While degrees in environmental, chemical, and bio-chemical engineering are common amongst workers in the wastewater engineering industry, a vast majority of wastewater engineers have their degree in civil engineering. Additionally, some states will require wastewater engineers to obtain a certificate and license through the state's board of civil engineering via the American Society of Civil Engineers.

A wastewater engineer's nature of work revolves around the design and management of water supply construction, water treatment processes, and sewage systems. However, their work can be much more versatile. For example, a wastewater engineer could specialize in providing solutions to protect towns from the dangers of flooding. A wastewater engineer may also be in charge of finding ideal areas to construct a municipal water treatment facility. On a more creative level, some wastewater engineers actually help to design water parks that exude fun and excitement and have pinnacle water conservation practices. Other roles of a wastewater engineer include:

- Design water storage tanks or facilities.
- Design pumping stations, systems, pipelines, and sewers for wastewater.
- Analyze storm water and floodplain drainage systems.
- Design domestic, industrial, and wastewater treatment plants.
- Conduct water-quality studies.
- Design a variety of plans for the development of new water resources.
- Design water runoff connection systems.

The list of roles and responsibilities continues on! To excel as a wastewater engineer, one must have an eye for detail, creativity, and innovation. By maintaining the water supply within the United States, wastewater engineers play a very crucial role in our society.

Sanitary engineering, also known as public health engineering or wastewater engineering, is the application of engineering methods to improve sanitation of human communities, primarily by providing the removal and disposal of human waste, and in addition to the supply of safe potable water. Traditionally a branch of civil engineering and now a subset of environmental engineering, in the mid-19th century, the discipline concentrated on the reduction of disease, then thought to be caused by miasma. This was accomplished mainly by the collection and segregation of sewerage flow in London specifically, and Great Britain generally. These and later regulatory improvements were reported in the United States as early as 1865.

It is also concerned with environmental factors that do not have an immediate and clearly understood effect on public health. Areas outside the purview of sanitary engineering include aesthetic concerns such as landscaping, and environmental conservation as it pertains to plants and animals.

Skills within this field are usually employed for the primary goal of disease prevention within human beings by assuring a supply of healthy drinking water, treatment of waste water, and removal of garbage from inhabited areas.

Compared to electrical engineering or mechanical engineering which are concerned primarily with closed systems, sanitary engineering is a very interdisciplinary field which may involve such elements as plumbing, fire protection, hydraulics, life safety, constructive modelling, information technology, project design, microbiology, pathology and the many divisions within environmental science and environmental technology. In some cases, considerations that fall within the field of social sciences and urban planning must be factored in as well.

Although sanitary engineering may be most associated with the design of sewers, sewage treatment and wastewater treatment facilities, recycling centers, public landfills and other things which are constructed, the term applies equally to a plan of action to reverse the effects of water pollution or soil contamination in a specific area.

1.3 HISTORY OF WATER SUPPLY AND SANITATION

Wastewater is water generated after the use of freshwater, raw water, drinking water or saline water in a variety of deliberate applications or processes. Another definition for wastewater is "Used water from any combination of domestic, industrial, commercial or agricultural activities, surface runoff / storm water, and any sewer inflow or sewer infiltration". In everyday usage, wastewater is commonly a synonym for sewage (also called sewerage, domestic wastewater, or municipal wastewater), which is wastewater that is produced by a community of people.

As a generic term wastewater may also be used to describe water containing contaminants accumulated in other settings, such as:

- **Industrial wastewater:** waterborne waste generated from a variety of industrial processes, such as manufacturing operations, mineral extraction, power generation, or water and wastewater treatment.
    - o Cooling water, released with potential thermal pollution after use to condense steam or reduce machinery temperatures by conduction or evaporation.
- **Leachate:** precipitation containing pollutants dissolved while percolating through ores, raw materials, products, or solid waste.
- **Return flow:** the flow of water carrying suspended soil, pesticide residues, or dissolved minerals and nutrients from irrigated cropland.

- **Surface runoff:** the flow of water occurring on the ground surface when excess rainwater, stormwater, meltwater, or other sources, can no longer sufficiently rapidly infiltrate in the soil.
    - Urban runoff, including water used for outdoor cleaning activity and landscape irrigation in densely populated areas created by urbanization.
- **Agricultural wastewater:** animal husbandry wastewater generated from confined animal operations.

We consider wastewater treatment as a water use because it is so interconnected with the other uses of water. Much of the water used by homes, industries, and businesses must be treated before it is released back to the environment.

If the term "wastewater treatment" is confusing to you, you might think of it as "sewage treatment." Nature has an amazing ability to cope with small amounts of water wastes and pollution, but it would be overwhelmed if we didn't treat the billions of gallons of wastewater and sewage produced every day before releasing it back to the environment. Treatment plants reduce pollutants in wastewater to a level nature can handle.

Wastewater also includes storm runoff. Although some people assume that the rain that runs down the street during a storm is fairly clean, it isn't. Harmful substances that wash off roads, parking lots, and rooftops can harm our rivers and lakes.

The history of water supply and sanitation is one of a logistical challenge to provide clean water and sanitation systems since the dawn of civilization. Where water resources, infrastructure or sanitation systems were insufficient, diseases spread and people fell sick or died prematurely.

Major human settlements could initially develop only where fresh surface water was plentiful, such as near rivers or natural springs. Throughout history, people have devised systems to make getting water into their communities and households and disposing of (and later also treating) wastewater more convenient.

The historical focus of sewage treatment was on the conveyance of raw sewage to a natural body of water, e.g. a river or ocean, where it would be diluted and dissipated. Early human habitations were often built next to water sources. Rivers would often serve as a crude form of natural sewage disposal.

Over the millennia, technology has dramatically increased the distances across which water can be relocated. Furthermore, treatment processes to purify drinking water and to treat wastewater have been improved.

1.3.1 Prehistory

During the Neolithic era, humans dug the first permanent water wells, from where vessels could be filled and carried by hand. Wells dug around 6500 BC have been found in the

Jezreel Valley. The size of human settlements was largely dependent on nearby available water.

Fig. 1. Skara Brae, a Neolithic village in Orkney, Scotland with home furnishings including water-flushing toilets 3180 BC–2500 BC.

A primitive indoor, tree bark lined, two-channel, stone, fresh and wastewater system appears to have featured in the houses of Skara Brae, and the Barnhouse Settlement, from around 3000 BCE, along with a cell-like enclave in a number of houses, of Skara Brae, that it has been suggested may have functioned as an early indoor latrine.

1.4 WASTEWATER REUSE ACTIVITIES

Wastewater reuse is an ancient practice, which has been applied since the dawn of human history, and is connected to the development of sanitation provision. Reuse of untreated municipal wastewater has been practiced for many centuries with the objective of diverting human waste outside of urban settlements. Likewise, land application of domestic wastewater is an old and common practice, which has gone through different stages of development.

Domestic wastewater was used for irrigation by prehistoric civilizations (e.g. Mesopotamian, Indus valley, and Minoan) since the Bronze Age (ca. 3200-1100 BC). Thereafter, wastewater was used for disposal, irrigation, and fertilization purposes by Hellenic civilizations and later by Romans in areas surrounding cities (e.g. Athens and Rome).

1.4.1 Ancient Americas

In ancient Peru, the Nazca people employed a system of interconnected wells and an underground watercourse known as puquios.

1.4.2 Mesopotamia

The Mesopotamians introduced the world to clay sewer pipes around 4000 BCE, with the earliest examples found in the Temple of Bel at Nippur and at Eshnunna, utilised to remove wastewater from sites, and capture rainwater, in wells. The city of Uruk also demonstrates the first examples of brick constructed latrines, from 3200 BCE. Clay pipes were later used in the Hittite city of Hattusa. They had easily detachable and replaceable segments, and allowed for cleaning.

1.4.3 Ancient Persia

The first sanitation systems within prehistoric Iran were built near the city of Zabol. Persian qanats and ab anbars have been used for water supply and cooling.

1.4.4 Ancient Egypt

The c. 2400 BCE, Pyramid of Sahure, and adjoining temple complex at Abusir, was discovered to have a network of copper drainage pipes.

1.4.5 Ancient China

Some of the earliest evidence of water wells are located in China. The Neolithic Chinese discovered and made extensive use of deep drilled groundwater for drinking. The Chinese text The Book of Changes, originally a divination text of the Western Zhou dynasty (1046 -771 BC), contains an entry describing how the ancient Chinese maintained their wells and protected their sources of water. Archaeological evidence and old Chinese documents reveal that the prehistoric and ancient Chinese had the aptitude and skills for digging deep water wells for drinking water as early as 6000 to 7000 years ago. A well excavated at the Hemedu excavation site was believed to have been built during the Neolithic era. The well was caused by four rows of logs with a square frame attached to them at the top of the well. Sixty additional tile wells southwest of Beijing are also believed to have been built around 600 BC for drinking and irrigation. Plumbing is also known to have been used in East Asia since the Qin and Han Dynasties of China.

Fig. 2. A Chinese ceramic model of a well with a water pulley system, excavated from a tomb of the Han Dynasty (202 BC - 220 AD) period.

1.5 INDUS VALLEY CIVILIZATION

The Indus Valley Civilization in Asia shows early evidence of public water supply and sanitation. The system the Indus developed and managed included a number of advanced features. An exceptional example is the Indus city of Lothal (c. 2350-1810 BCE). In Lothal the ruler's house had their own private bathing platform and latrine, which was connected to an open street drain that discharged into the towns dock. A number of the other houses of the acropolis had burnished brick bathing platforms, that drained into a covered brick sewer, held together with a gypsum-based mortar, that ran to a soak pit outside the towns walls, while the lower town offered soak jars (large buried urns, with a hole in the bottom to permit liquids to drain), the latter of which were regularly emptied and cleaned. Water was supplied from two wells in the town, one in the acropolis, and the other on the edge of the dock.

The urban areas of the Indus Valley civilization included public and private baths. Sewage was disposed through underground drains built with precisely laid bricks, and a sophisticated water management system with numerous reservoirs was established. In the drainage systems, drains from houses were connected to wider public drains. Many of the buildings at Mohenjo-daro had two or more stories. Water from the roof and upper storey bathrooms was carried through enclosed terracotta pipes or open chutes that emptied out onto the street drains.

Fig. 3. A large well and bathing platforms at Harappa, remains of the city's final phase of occupation from 2200 to 1900 BC.

Fig. 4. The bathroom-toilet structure of the ruler's house, on Lothal's acropolis! c.2350 BCE.

Fig. 5. Bathing platform and communal drain, Lothal's acropolis! c.2350 BCE.

Fig. 6. Well, and drain, Lothal's acropolis! c.2350 BCE.

The earliest evidence of urban sanitation was seen in Harappa, Mohenjo-daro, and the recently discovered Rakhigarhi of Indus Valley civilization. This urban plan included the world's first urban sanitation systems. Within the city, individual homes or groups of homes obtained water from wells. From a room that appears to have been set aside for bathing, waste water was directed to covered drains, which lined the major streets.

Devices such as shadoofs were used to lift water to ground level. Ruins from the Indus Valley Civilization like Mohenjo-daro in Pakistan and Dholavira in Gujarat in India had settlements with some of the ancient world's most sophisticated sewage systems. They included drainage channels, rainwater harvesting, and street ducts.

Stepwells have mainly been used in the Indian subcontinent.

1.6 ANCIENT MEDITERRANEAN

1.6.1 Ancient Greece

The ancient Greek civilization of Crete, known as the Minoan civilization, was the first civilization to use underground clay pipes for sanitation and water supply. Their capital, Knossos, had a well-organized water system for bringing in clean water, taking out waste water and storm sewage canals for overflow when there was heavy rain. It was also one of the first uses of a flush toilet, dating back to the 18th century BC. The Minoan civilization had stone sewers that were periodically flushed with clean water. In addition to sophisticated water and sewer systems they devised elaborate heating systems. The Ancient Greeks of Athens and Asia Minor also used an indoor plumbing system, used for pressurized showers. The Greek inventor Heron used pressurized piping for fire fighting purposes in the City of Alexandria. The Mayans were the third earliest civilization to have employed a system of indoor plumbing using pressurized water.

An inverted siphon system, along with glass covered clay pipes, was used for the first time in the palaces of Crete, Greece. It is still in working condition, after about 3000 years.

1.6.2 Roman Empire

Fig. 7. Pont du Gard, a Roman aqueduct in France.

In ancient Rome, the **Cloaca Maxima,** considered a marvel of engineering, discharged into the Tiber. Public latrines were built over the Cloaca Maxima.

Beginning in the Roman era a water wheel device known as a noria supplied water to aqueducts and other water distribution systems in major cities in Europe and the Middle East.

The Roman Empire had indoor plumbing, meaning a system of aqueducts and pipes that terminated in homes and at public wells and fountains for people to use. Rome and other nations used lead pipes; while commonly thought to be the cause of lead poisoning in the Roman Empire, the combination of running water which did not stay in contact with the pipe for long and the deposition of precipitation scale actually mitigated the risk from lead pipes.

Roman towns and garrisons in the United Kingdom between 46 BC and 400 AD had complex sewer networks sometimes constructed out of hollowed-out elm logs, which were shaped so that they butted together with the down-stream pipe providing a socket for the upstream pipe.

1.7 MEDIEVAL AND EARLY MODERN AGES

1.7.1 Nepal

Fig. 8. People waiting in line for water from Manga Hiti in the city of Patan, Nepal.

In Nepal the construction of water conduits like drinking fountains and wells is considered a pious act.

A drinking water supply system was developed starting at least as early as 550 AD. This dhunge dhara or hiti system consists of carved stone fountains through which water flows uninterrupted from underground sources. These are supported by numerous ponds and canals that form an elaborate network of water bodies, created as a water resource during the dry season and to help alleviate the water pressure caused by the monsoon rains. After the introduction of modern, piped water systems, starting in the late 19th century, this old system has fallen into disrepair and some parts of it are lost forever. Nevertheless, many people of Nepal still rely on the old hitis on a daily basis.

In 2008 the dhunge dharas of the Kathmandu Valley produced 2.95 million litres of water per day.

Of the 389 stone spouts found in the Kathmandu Valley in 2010, 233 were still in use, serving about 10% of Kathmandu's population. 68 had gone dry, 45 were lost entirely and 43 were connected to the municipal water supply instead of their original source.

1.7.2 Islamic World

Fig. 9. The Comares Baths at the Alhambra in Granada, Spain (14th century).

Islam stresses the importance of cleanliness and personal hygiene. Islamic hygienical jurisprudence, which dates back to the 7th century, has a number of elaborate rules. Taharah (ritual purity) involves performing wudu (ablution) for the five daily salah (prayers), as well as regularly performing ghusl (bathing), which led to bathhouses being built across the Islamic world. Islamic toilet hygiene also requires washing with water after using the toilet, for purity and to minimize germs.

In the Abbasid Caliphate (8th-13th centuries), its capital city of Baghdad (Iraq) had 65,000 baths, along with a sewer system. Cities of the medieval Islamic world had water supply systems powered by hydraulic technology that supplied drinking water along with much greater quantities of water for ritual washing, mainly in mosques and hammams (baths). Bathing establishments in various cities were rated by Arabic writers in travel guides. Medieval Islamic cities such as Baghdad, Córdoba (Islamic Spain), Fez (Morocco) and Fustat (Egypt) also had sophisticated waste disposal and sewage systems with interconnected networks of sewers. The city of Fustat also had multi-storey tenement buildings (with up to six floors) with flush toilets, which were connected to a water supply system, and flues on each floor carrying waste to underground channels.

Al-Karaji (c. 953–1029) wrote a book, The Extraction of Hidden Waters, which presented ground-breaking ideas and descriptions of hydrological and hydrogeological perceptions such as components of the hydrological cycle, groundwater quality, and driving factors of groundwater flow. He also gave an early description of a water filtration process.

1.8 POST-CLASSICAL EAST AFRICA

In post-classical Kilwa plumbing was prevalent in the stone homes of the natives. The Husani Kubwa Palace as well as other buildings for the ruling elite and wealthy included the luxury of indoor plumbing.

1.8.1 Medieval Europe

Fig. 10. Agkistro Byzantine bath.

Fig. 11. A 1939 conceptual illustration showing various ways that typhoid bacteria can contaminate a water well (center).

Fig. 12. Waterworks (Wasserkunst) and fountain from 1602 in Wismar, Germany.

Christianity places an emphasis on hygiene. Despite the denunciation of the mixed bathing style of Roman pools by early Christian clergy, as well as the pagan custom of women naked bathing in front of men, this did not stop the Church from urging its followers to go to public baths for bathing, which contributed to hygiene and good health according to the Church Fathers, Clement of Alexandria and Tertullian. The Church built public bathing facilities that were separate for both sexes near monasteries and pilgrimage sites; also, the popes situated baths within church basilicas and monasteries since the early Middle Ages. Pope Gregory the Great urged his followers on value of bathing as a bodily need.

Contrary to popular belief bathing and sanitation were not lost in Europe with the collapse of the Roman Empire. Public bathhouses were common in medieval Christendom larger towns and cities such as Constantinople, Paris, Regensburg, Rome and Naples. And great bathhouses were built in Byzantine centers such as Constantinople and Antioch.

There is little record of other sanitation systems (apart of sanitation in ancient Rome) in most of Europe until the High Middle Ages. Unsanitary conditions and overcrowding were widespread throughout Europe and Asia during the Middle Ages. This resulted in pandemics such as the Plague of Justinian (541–542) and the Black Death (1347–1351), which killed tens of millions of people. Very high infant and child mortality prevailed in Europe throughout medieval times, due partly to deficiencies in sanitation.

In medieval European cities, small natural waterways used for carrying off wastewater were eventually covered over and functioned as sewers. London's River Fleet is such a system. Open drains, or gutters, for waste water run-off ran along the center of some streets. These were known as "kennels" (i.e., canals, channels), and in Paris were sometimes known as "split streets," as the waste water running along the middle physically split the streets into two halves. The first closed sewer constructed in Paris was designed by Hugues Aubird in 1370 on Rue Montmartre (Montmartre Street), and was 300 meters long. The original purpose of designing and constructing a closed sewer in Paris was less-so for waste management as much as it was to hold back the stench coming from the odorous waste water. In Dubrovnik, then known as Ragusa (Latin name), the Statute of 1272 set out the parameters for the construction of septic tanks and channels for the removal of dirty water. Throughout the 14th and 15th century the sewage system was built, and it is still operational today, with minor changes and repairs done in recent centuries. Pail closets, outhouses, and cesspits were used to collect human waste. The use of human waste as fertilizer was especially important in China and Japan, where cattle manure was less available. However, most cities did not have a functioning sewer system before the Industrial era , relying instead on nearby rivers or occasional rain showers to wash away the sewage from the streets . In some places, waste water simply ran down the streets, which had stepping stones to keep pedestrians out of the muck, and eventually drained as runoff into the local watershed.

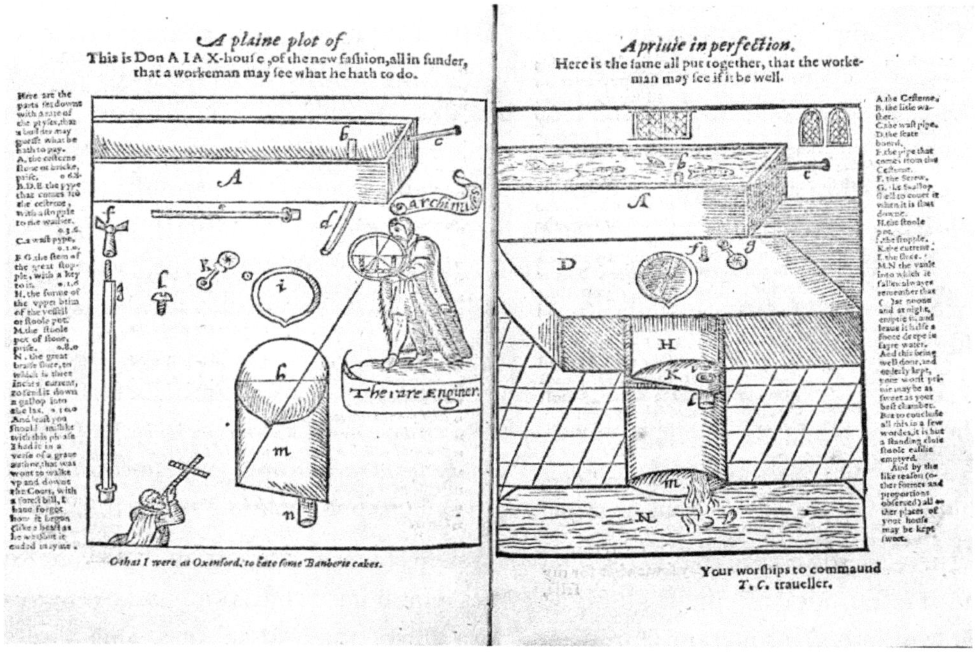

Fig. 13. John Harington's toilet.

In the 16th century, Sir John Harington invented a flush toilet as a device for Queen Elizabeth I (his godmother) that released wastes into cesspools.

After the adoption of gunpowder, municipal outhouses became an important source of raw material for the making of saltpeter in European countries.

In London, the contents of the city's outhouses were collected every night by commissioned wagons and delivered to the nitrite beds where it was laid into specially designed soil beds to produce earth rich in mineral nitrates. The nitrate rich-earth would be then further processed to produce saltpeter, or potassium nitrate, an important ingredient in black powder that played a part in the making of gunpowder.

1.9 CLASSIC AND EARLY MODERN MESOAMERICA

The Classic Maya at Palenque had underground aqueducts and flush toilets; the Classic Maya even used household water filters using locally abundant limestone carved into a porous cylinder, made so as to work in a manner strikingly similar to Modern ceramic water filters.

In Spain and Spanish America, a community operated watercourse known as an acequia, combined with a simple sand filtration system, provided potable water.

1.9.1 Sewage Farms for Disposal and Irrigation

"Sewage farms" (i.e. wastewater application to the land for disposal and agricultural use) were operated in Bunzlau (Silesia) in 1531, in Edinburgh (Scotland) in 1650, in Paris (France) in 1868, in Berlin (Germany) in 1876 and in different parts of the USA since 1871, where wastewater was used for beneficial crop production. In the following centuries (16th and 18th centuries) in many rapidly growing countries/cities of Europe (e.g. Germany, France) and the United States, "sewage farms" were increasingly seen as a solution for the disposal of large volumes of the wastewater, some of which are still in operation today. Irrigation with sewage and other wastewater effluents has a long history also in China and India; while also a large "sewage farm" was established in Melbourne, Australia, in 1897.

1.10 MODERN AGE

1.10.1 Sewer Systems

Fig. 14. Many industrialized cities had incomplete public sanitation well into the 20th century. Outhouses in Brisbane, Australia, around 1950.

A significant development was the construction of a network of sewers to collect wastewater. In some cities, including Rome, Istanbul (Constantinople) and Fustat, networked ancient sewer systems continue to function today as collection systems for those cities' modernized sewer systems. Instead of flowing to a river or the sea, the pipes have been re-routed to modern sewer treatment facilities.

Basic sewer systems were used for waste removal in ancient Mesopotamia, where vertical shafts carried the waste away into cesspools. Similar systems existed in the Indus Valley civilization in modern-day India and in Ancient Crete and Greece. In the Middle Ages the sewer systems built by the Romans fell into disuse and waste was collected into cesspools that were periodically emptied by workers known as 'rakers' who would often sell it as fertilizer to farmers outside the city.

Archaeological discoveries have shown that some of the earliest sewer systems were developed in the third millennium BCE in the ancient cities of Harappa and Mohenjo-daro in present-day Pakistan. The primitive sewers were carved in the ground alongside buildings. This discovery reveals the conceptual understanding of waste disposal by early civilizations.

However, until the Enlightenment era, little progress was made in water supply and sanitation and the engineering skills of the Romans were largely neglected throughout Europe. This began to change in the 17th and 18th centuries with a rapid expansion in waterworks and pumping systems.

The tremendous growth of cities in Europe and North America during the Industrial Revolution quickly led to crowding, which acted as a constant source for the outbreak of disease. :4–8 As cities grew in the 19th century concerns were raised about public health. :33–62 As part of a trend of municipal sanitation programs in the late 19th and 20th centuries, many cities constructed extensive gravity sewer systems to help control outbreaks of disease such as typhoid and cholera. :29–34 Storm and sanitary sewers were necessarily developed along with the growth of cities. By the 1840s the luxury of indoor plumbing, which mixes human waste with water and flushes it away, eliminated the need for cesspools.

Modern sewerage systems were first built in the mid-nineteenth century as a reaction to the exacerbation of sanitary conditions brought on by heavy industrialization and urbanization. Baldwin Latham, a British civil engineer contributed to the rationalization of sewerage and house drainage systems and was a pioneer in sanitary engineering. He developed the concept of oval sewage pipe to facilitate sewer drainage and to prevent sludge deposition and flooding. Due to the contaminated water supply, cholera outbreaks occurred in 1832, 1849 and 1855 in London, killing tens of thousands of people. This, combined with the Great Stink of 1858, when the smell of untreated human waste in the River Thames became overpowering, and the report into sanitation reform of the

Royal Commissioner Edwin Chadwick, led to the Metropolitan Commission of Sewers appointing Joseph Bazalgette to construct a vast underground sewage system for the safe removal of waste. Contrary to Chadwick's recommendations, Bazalgette's system, and others later built in Continental Europe, did not pump the sewage onto farm land for use as fertilizer; it was simply piped to a natural waterway away from population centres, and pumped back into the environment.

1.10.2 Liverpool, London and Other Cities in the UK

As recently as the late 19th-century sewerage systems in some parts of the rapidly industrializing United Kingdom were so inadequate that water-borne diseases such as cholera and typhoid remained a risk.

From as early as 1535 there were efforts to stop polluting the River Thames in London. Beginning with an Act passed that year that was to prohibit the dumping of excrement into the river. Leading up to the Industrial Revolution the River Thames was identified as being thick and black due to sewage, and it was even said that the river "smells like death." As Britain was the first country to industrialize, it was also the first to experience the disastrous consequences of major urbanization and was the first to construct a modern sewerage system to mitigate the resultant unsanitary conditions. During the early 19th century, the River Thames was effectively an open sewer, leading to frequent outbreaks of cholera epidemics. Proposals to modernize the sewerage system had been made during 1856 but were neglected due to lack of funds. However, after the Great Stink of 1858, Parliament realized the urgency of the problem and resolved to create a modern sewerage system.

However, ten years earlier and 200 miles to the north, James Newlands, a Scottish Engineer, was one of a celebrated trio of pioneering officers appointed under a private Act, the Liverpool Sanitory Act by the Borough of Liverpool Health of Towns Committee. The other officers appointed under the Act were William Henry Duncan, Medical Officer for Health, and Thomas Fresh, Inspector of Nuisances (an early antecedent of the environmental health officer). One of five applicants for the post, Newlands was appointed Borough Engineer of Liverpool on 26 January 1847.

He made a careful and exact survey of Liverpool and its surroundings, involving approximately 3,000 geodetical observations, and resulting in the construction of a contour map of the town and its neighbourhood, on a scale of one inch to 20 feet (6.1 m). From this elaborate survey Newlands proceeded to lay down a comprehensive system of outlet and contributory sewers, and main and subsidiary drains, to an aggregate extent of nearly 300 miles (480 km). The details of this projected system he presented to the Corporation in April 1848.

In July 1848 James Newlands' sewer construction programme began, and over the next 11 years 86 miles (138 km) of new sewers were built. Between 1856 and 1862 another

58 miles (93 km) were added. This programme was completed in 1869. Before the sewers were built, life expectancy in Liverpool was 19 years, and by the time Newlands retired it had more than doubled.

Joseph Bazalgette, a civil engineer and Chief Engineer of the Metropolitan Board of Works, was given responsibility for the work. He designed an extensive underground sewerage system that diverted waste to the Thames Estuary, downstream of the main center of population. Six main interceptor sewers, totaling almost 100 miles (160 km) in length, were constructed, some incorporating stretches of London's 'lost' rivers. Three of these sewers were north of the river, the southernmost, low-level one being incorporated in the Thames Embankment. The Embankment also allowed new roads, new public gardens, and the Circle Line of the London Underground.

The intercepting sewers, constructed between 1859 and 1865, were fed by 450 miles (720 km) of main sewers that, in turn, conveyed the contents of some 13,000 miles (21,000 km) of smaller local sewers. Construction of the interceptor system required 318 million bricks, 2.7 million cubic metres of excavated earth and 670,000 cubic metres of concrete. Gravity allowed the sewage to flow eastwards, but in places such as Chelsea, Deptford and Abbey Mills, pumping stations were built to raise the water and provide sufficient flow. Sewers north of the Thames feed into the Northern Outfall Sewer, which fed into a major treatment works at Beckton. South of the river, the Southern Outfall Sewer extended to a similar facility at Crossness. With only minor modifications, Bazalgette's engineering achievement remains the basis for sewerage design up into the present day.

In Merthyr Tydfil, a large town in South Wales, most houses discharged their sewage to individual cess-pits which persistently overflowed causing the pavements to be awash with foul sewage.

1.10.3 Paris, France

In 1802, Napoleon built the Ourcq canal which brought 70,000 cubic meters of water a day to Paris, while the Seine river received up to 100,000 cubic meters (3,500,000 cu ft) of wastewater per day. The Paris cholera epidemic of 1832 sharpened the public awareness of the necessity for some sort of drainage system to deal with sewage and wastewater in a better and healthier way. Between 1865 and 1920 Eugene Belgrand lead the development of a large scale system for water supply and wastewater management. Between these years approximately 600 kilometers of aqueducts were built to bring in potable spring water, which freed the poor quality water to be used for flushing streets and sewers. By 1894 laws were passed which made drainage mandatory. The treatment of Paris sewage, though, was left to natural devices as 5,000 hectares of land were used to spread the waste out to be naturally purified. Further, the lack of sewage treatment left Parisian sewage pollution to become concentrated downstream in the town of Clichy, effectively forcing residents to pack up and move elsewhere.

The 19th century brick-vaulted Paris sewers serve as a tourist attraction nowadays.

1.10.4 Hamburg and Frankfurt, Germany

The first comprehensive sewer system in a German city was built in Hamburg, Germany, in the mid-19th century.

In 1863, work began on the construction of a modern sewerage system for the rapidly growing city of Frankfurt am Main, based on design work by William Lindley. 20 years after the system's completion, the death rate from typhoid had fallen from 80 to 10 per 100,000 inhabitants.

1.10.5 United States

The first sewer systems in the United States were built in the late 1850s in Chicago and Brooklyn.

In the United States, the first sewage treatment plant using chemical precipitation was built in Worcester, Massachusetts, in 1890.

1.11 SEWAGE TREATMENT

Initially the gravity sewer systems discharged sewage directly to surface waters without treatment. Later, cities attempted to treat the sewage before discharge in order to prevent water pollution and waterborne diseases. During the half-century around 1900, these public health interventions succeeded in drastically reducing the incidence of water-borne diseases among the urban population, and were an important cause in the increases of life expectancy experienced at the time.

1.12 APPLICATION ON AGRICULTURAL LAND

Early techniques for sewage treatment involved land application of sewage on agricultural land. One of the first attempts at diverting sewage for use as a fertilizer in the farm was made by the cotton mill owner James Smith in the 1840s. He experimented with a piped distribution system initially proposed by James Vetch that collected sewage from his factory and pumped it into the outlying farms, and his success was enthusiastically followed by Edwin Chadwick and supported by organic chemist Justus von Liebig.

The idea was officially adopted by the Health of Towns Commission, and various schemes (known as sewage farms) were trialled by different municipalities over the next 50 years. At first, the heavier solids were channeled into ditches on the side of the farm and were covered over when full, but soon flat-bottomed tanks were employed as reservoirs for the sewage; the earliest patent was taken out by William Higgs in 1846 for "tanks or reservoirs in which the contents of sewers and drains from cities, towns and villages are to be collected and the solid animal or vegetable matters therein contained, solidified and dried..." Improvements to the design of the tanks included the introduction of the horizontal-flow tank in the 1850s and the radial-flow tank in 1905. These tanks had to be manually de-sludged periodically, until the introduction of automatic mechanical de-sludgers in the early 1900s.

1.13 CHEMICAL TREATMENT AND SEDIMENTATION

As pollution of water bodies became a concern, cities attempted to treat the sewage before discharge. In the late 19th century some cities began to add chemical treatment and sedimentation systems to their sewers. During the half-century around 1900, these public health interventions succeeded in drastically reducing the incidence of water-borne diseases among the urban population, and were an important cause in the increases of life expectancy experienced at the time.

Odor was considered the big problem in waste disposal and to address it, sewage could be drained to a lagoon, or "settled" and the solids removed, to be disposed of separately. This process is now called "primary treatment" and the settled solids are called "sludge." At the end of the 19th century, since primary treatment still left odor problems, it was discovered that bad odors could be prevented by introducing oxygen into the decomposing sewage. This was the beginning of the biological aerobic and anaerobic treatments which are fundamental to wastewater processes.

The precursor to the modern septic tank was the cesspool in which the water was sealed off to prevent contamination and the solid waste was slowly liquified due to anaerobic action; it was invented by L.H Mouras in France in the 1860s. Donald Cameron, as City Surveyor for Exeter patented an improved version in 1895, which he called a 'septic tank'; septic having the meaning of 'bacterial'. These are still in worldwide use, especially in rural areas unconnected to large-scale sewage systems.

1.14 BIOLOGICAL TREATMENT

It was not until the late 19th century that it became possible to treat the sewage by biologically decomposing the organic components through the use of microorganisms and removing the pollutants. Land treatment was also steadily becoming less feasible, as cities grew and the volume of sewage produced could no longer be absorbed by the farmland on the outskirts.

Edward Frankland conducted experiments at the sewage farm in Croydon, England, during the 1870s and was able to demonstrate that filtration of sewage through porous gravel produced a nitrified effluent (the ammonia was converted into nitrate) and that the filter remained unclogged over long periods of time. This established the then revolutionary possibility of biological treatment of sewage using a contact bed to oxidize the waste. This concept was taken up by the chief chemist for the London Metropolitan Board of Works, William Libdin, in 1887:

In all probability the true way of purifying sewage will be first to separate the sludge, and then turn into neutral effluent... retain it for a sufficient period, during which time it should be fully aerated, and finally discharge it into the stream in a purified condition. This is indeed what is aimed at and imperfectly accomplished on a sewage farm.

From 1885 to 1891 filters working on this principle were constructed throughout the UK and the idea was also taken up in the US at the Lawrence Experiment Station in Massachusetts, where Frankland's work was confirmed. In 1890 the LES developed a 'trickling filter' that gave a much more reliable performance.

Contact beds were developed in Salford, Lancashire and by scientists working for the London City Council in the early 1890s. According to Christopher Hamlin, this was part of a conceptual revolution that replaced the philosophy that saw "sewage purification as the prevention of decomposition with one that tried to facilitate the biological process that destroy sewage naturally."

Contact beds were tanks containing an inert substance, such as stones or slate, that maximized the surface area available for the microbial growth to break down the sewage. The sewage was held in the tank until it was fully decomposed and it was then filtered out into the ground. This method quickly became widespread, especially in the UK, where it was used in Leicester, Sheffield, Manchester and Leeds. The bacterial bed was simultaneously developed by Joseph Corbett as Borough Engineer in Salford and experiments in 1905 showed that his method was superior in that greater volumes of sewage could be purified better for longer periods of time than could be achieved by the contact bed.

The Royal Commission on Sewage Disposal published its eighth report in 1912 that set what became the international standard for sewage discharge into rivers; the '20:30 standard', which allowed "2 parts per hundred thousand" of Biochemical oxygen demand and "3 parts per hundred thousand" of suspended solid.

1.15 ACTIVATED SLUDGE PROCESS

Most cities in the Western world added more expensive systems for sewage treatment in the early 20th century, after scientists at the University of Manchester discovered the sewage treatment process of activated sludge in 1912.

This Section is An Excerpt From Activated Sludge History. The activated sludge process was discovered in 1913 in the United Kingdom by two engineers, Edward Ardern and W.T. Lockett, who were conducting research for the Manchester Corporation Rivers Department at Davyhulme Sewage Works. In 1912, Dr. Gilbert Fowler, a scientist at the University of Manchester, observed experiments being conducted at the Lawrence Experiment Station at Massachusetts involving the aeration of sewage in a bottle that had been coated with algae. Fowler's engineering colleagues, Ardern and Lockett, experimented on treating sewage in a draw-and-fill reactor, which produced a highly treated effluent. They aerated the waste-water continuously for about a month and were able to achieve a complete nitrification of the sample material. Believing that the sludge had been activated (in a similar manner to activated carbon) the process was named activated sludge. Not until much later was it realized that what had actually occurred was a means to concentrate biological organisms, decoupling the liquid retention time (ideally, low, for a compact treatment system) from the solids retention time (ideally, fairly high, for an effluent low in BOD5 and ammonia.)

Their results were published in their seminal 1914 paper, and the first full-scale continuous-flow system was installed at Worcester two years later. In the aftermath of the First World War the new treatment method spread rapidly, especially to the USA, Denmark, Germany and Canada. By the late 1930s, the activated sludge treatment became a well-known biological wastewater treatment process in those countries where sewer systems and sewage treatment plants were common.

1.15.1 Toilets

With the onset of the industrial revolution and related advances in technology, the flush toilet began to emerge into its modern form. It needs to be connected to a sewer system though. Where this is not feasible or desired, dry toilets are an alternative option.

1.15.2 Water Supply

An ambitious engineering project to bring fresh water from Hertfordshire to London was undertaken by Hugh Myddleton, who oversaw the construction of the New River between 1609 and 1613. The New River Company became one of the largest private water companies of the time, supplying the City of London and other central areas. The first civic system of piped water in England was established in Derby in 1692, using wooden pipes, which was common for several centuries. The Derby Waterworks included waterwheel-powered pumps for raising water out of the River Derwent and storage tanks for distribution.

It was in the 18th century that a rapidly growing population fueled a boom in the establishment of private water supply networks in London. The Chelsea Waterworks Company was established in 1723 "for the better supplying the City and Liberties of Westminster and parts adjacent with water". The company created extensive ponds in the area bordering Chelsea and Pimlico using water from the tidal Thames. Other waterworks were established in London, including at West Ham in 1743, at Lea Bridge before 1767, Lambeth Waterworks Company in 1785, West Middlesex Waterworks Company in 1806 and Grand Junction Waterworks Company in 1811.

The S-bend pipe was invented by Alexander Cummings in 1775 but became known as the U-bend following the introduction of the U-shaped trap by Thomas Crapper in 1880. The first screw-down water tap was patented in 1845 by Guest and Chrimes, a brass foundry in Rotherham.

1.16 WATER TREATMENT

1.16.1 Sand Filter

Sir Francis Bacon attempted to desalinate sea water by passing the flow through a sand filter. Although his experiment did not succeed, it marked the beginning of a new interest in the field.

The first documented use of sand filters to purify the water supply dates to 1804, when the owner of a bleachery in Paisley, Scotland, John Gibb, installed an experimental filter, selling his unwanted surplus to the public. This method was refined in the following two decades by engineers working for private water companies, and it culminated in the first treated public water supply in the world, installed by engineer James Simpson for the Chelsea Waterworks Company in London in 1829. This installation provided filtered water for every resident of the area, and the network design was widely copied throughout the United Kingdom in the ensuing decades.

The Metropolis Water Act introduced the regulation of the water supply companies in London, including minimum standards of water quality for the first time. The Act "made provision for securing the supply to the Metropolis of pure and wholesome water", and required that all water be "effectually filtered" from 31 December 1855. This was followed up with legislation for the mandatory inspection of water quality, including comprehensive chemical analyses, in 1858. This legislation set a worldwide precedent for similar state public health interventions across Europe.The Metropolitan Commission of Sewers was formed at the same time, water filtration was adopted throughout the country, and new water intakes on the Thames were established above Teddington Lock. Automatic pressure filters, where the water is forced under pressure through the filtration system, were innovated in 1899 in England.

1.16.2 Water Chlorination

In what may have been one of the first attempts to use chlorine, William Soper used chlorinated lime to treat the sewage produced by typhoid patients in 1879.

In a paper published in 1894, Moritz Traube formally proposed the addition of chloride of lime (calcium hypochlorite) to water to render it "germ-free." Two other investigators confirmed Traube's findings and published their papers in 1895. Early attempts at implementing water chlorination at a water treatment plant were made in 1893 in Hamburg, Germany, and in 1897 the town of Maidstone, in Kent, England, was the first to have its entire water supply treated with chlorine.

Permanent water chlorination began in 1905, when a faulty slow sand filter and a contaminated water supplyled to a serious typhoid fever epidemic in Lincoln, England. Dr. Alexander Cruickshank Houston used chlorination of the water to stem the epidemic. His installation fed a concentrated solution of chloride of lime to the water being treated. The chlorination of the water supply helped stop the epidemic and as a precaution, the chlorination was continued until 1911 when a new water supply was instituted.

Manual Control Chlorinator for the liquefaction of chlorine for water purification, early 20th century. From Chlorination of Water by Joseph Race, 1918.

The first continuous use of chlorine in the United States for disinfection took place in 1908 at Boonton Reservoir (on the Rockaway River), which served as the supply for Jersey City, New Jersey. Chlorination was achieved by controlled additions of dilute solutions of chloride of lime (calcium hypochlorite) at doses of 0.2 to 0.35 ppm. The treatment process was conceived by Dr. John L. Leal and the chlorination plant was designed by George Warren Fuller. Over the next few years, chlorine disinfection using chloride of lime were rapidly installed in drinking water systems around the world.

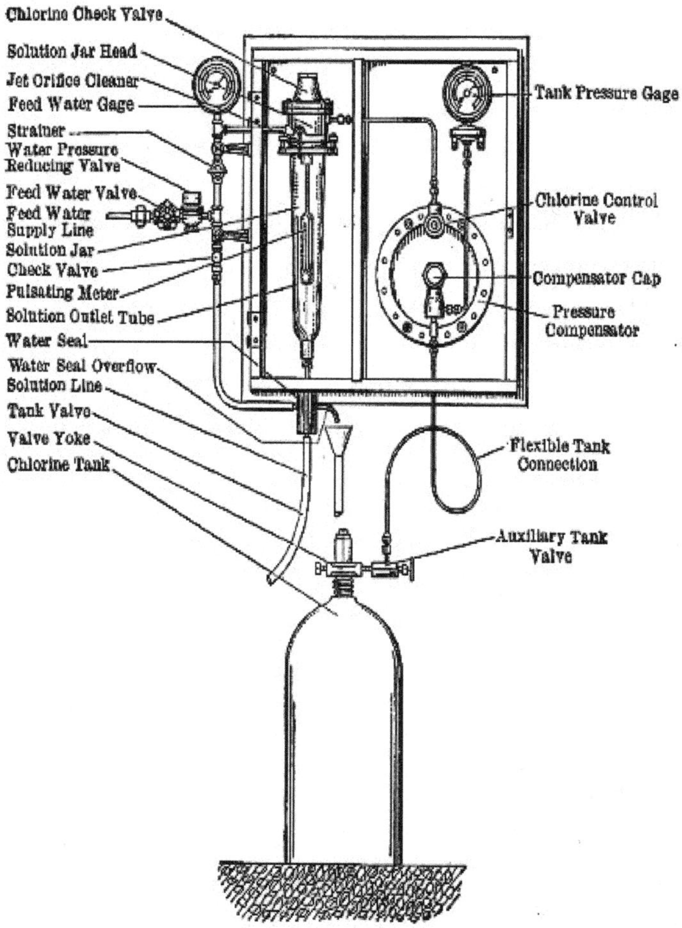

Fig. 15. Water chlorination.

The technique of purification of drinking water by use of compressed liquefied chlorine gas was developed by a British officer in the Indian Medical Service, Vincent B. Nesfield, in 1903. According to his own account, "It occurred to me that chlorine gas might be found satisfactory ... if suitable means could be found for using it.... The next important question was how to render the gas portable. This might be accomplished in two ways: By liquefying it, and storing it in lead-lined iron vessels, having a jet with a very fine capillary canal, and fitted with a tap or a screw cap. The tap is turned on, and the cylinder placed in the amount of water required. The chlorine bubbles out, and in ten to fifteen minutes the water is absolutely safe. This method would be of use on a large scale, as for service water carts."

U.S. Army Major Carl Rogers Darnall, Professor of Chemistry at the Army Medical School, gave the first practical demonstration of this in 1910. Shortly thereafter, Major William J. L. Lyster of the Army Medical Department used a solution of calcium hypochlorite in a linen bag to treat water. For many decades, Lyster's method remained the standard for U.S. ground forces in the field and in camps, implemented in the form of the familiar Lyster Bag (also spelled Lister Bag). This work became the basis for present day systems of municipal water purification.

1.16.3 Fluoridation

Water fluoridation is a practice adding fluoride to drinking water for the purpose of decreasing tooth decay.

The architect of these first fluoride studies was Dr. H. Trendley Dean, head of the Dental Hygiene Unit at the National Institute of Health (NIH). Dean began investigating the epidemiology of fluorosis in 1931y the late 1930s, he and his staff had made a critical discovery. Namely, fluoride levels of up to 1.0 ppm in drinking water did not cause enamel fluorosis in most people and only mild enamel fluorosis in a small percentage of people. Proof That Fluoride Prevents Caries This finding sent Dean's thoughts spiraling in a new direction. He recalled from reading McKay's and Black's studies on fluorosis that mottled tooth enamel is unusually resistant to decay. Dean wondered whether adding fluoride to drinking water at physically and cosmetically safe levels would help fight tooth decay. This hypothesis, Dean told his colleagues, would need to be tested.In 1944, Dean got his wish. That year, the City Commission of Grand Rapids, Michigan-after numerous discussions with researchers from the PHS, the Michigan Department of Health, and other public health organizations-voted to add fluoride to its public water supply the following year. In 1945, Grand Rapids became the first city in the world to fluoridate its drinking water.The Grand Rapids water fluoridation study was originally sponsored by the U.S. Surgeon General, but was taken over by the NIDR shortly after the Institute's inception in 1948.

1.16.4 Trends

The Sustainable Development Goal 6 formulated in 2015 includes targets on access to water supply and sanitation at a global level. In developing countries, self-supply of water and sanitation is used as an approach of incremental improvements to water and sanitation services, which are mainly financed by the user. Decentralized wastewater systems are also growing in importance to achieve sustainable sanitation.

1.17 TYPES OF WASTEWATER

Often used interchangeably with the term sewage, "sewage" technically denotes any wastewaters which pass through a sewer. Prior to entering a wastewater treatment plant, wastewater is sometimes called raw wastewater or raw sewage.

Domestic wastewater originates from activities such as restroom usage, bathing, food preparation and laundry. Commercial wastewater from non-domestic sources, such as beauty salons or auto body repair shops, for example. This wastewater may contain hazardous materials and requires special treatment or disposal. Industrial wastewater originates from industrial or commercial manufacturing processes, such as agriculture, and are usually more difficult to treat than domestic wastes. Industrial wastewater's composition varies on an industry-by-industry basis.

1.17.1 Organic Content of Wastewater

The organic content of wastewater is made up of human feces, protein, fat, vegetable and sugar material from food preparation, as well as soaps. Some of this organic content is dissolved into the water and some exist as separate particles. The portion of organic material that does not dissolve but remains suspended in the water is known as suspended solids. Wastewater is treated to remove as much organic material as possible.

1.17.2 Implications for Microorganisms

Naturally occurring soil and water bacteria eat the organic waste in wastewater and use it as a food and energy source to grow rapidly. In a natural water environment where there is plenty of oxygen dissolved in the water, aerobic bacteria eat the organic material and form a slime of new bacterial cells and dissolved salt-waste products.

If undiluted wastewater is left on its own, anaerobic bacteria decompose the waste organic material and release odorous gases such as hydrogen sulphide. Odor-free gases such as methane and carbon dioxide can also be released.

Where there is an overwhelming amount of wastewater, all the oxygen will be used up and the anaerobic bacteria will take over, making the water go septic. This is ultimately harmful to fish and other forms of life dependent on oxygen, on occasion creating dead zones.

1.17.3 Inorganic Matter

Inorganic minerals, metals and compounds, such as sodium, copper, lead and zinc are common in wastewater from both sewage and wastewater. They can originate from industrial and commercial sources, stormwater, and inflow and infiltration from cracked pipes. Most inorganic substances are stable and cannot be broken down easily by organisms in wastewater.

1.17.4 Nutrients

Excessive nutrients such as phosphorus and nitrogen can cause eutrophication, which can also be toxic to aquatic organisms. This also promotes excessive plant growth and reduces oxygen availability, altering habitats and potentially endangering certain species.

1.17.5 Other Wastewater Pollutants

Bacteria, viruses and disease-causing pathogens in wastewater can pollute beaches and contaminate shellfish populations. Fecal coliform bacteria in human waste is typically harmless, but there are pathogens that can negatively impact human health. These can be bacteria such as typhoid or viruses such as hepatitis B, for example. Direct contact with these pathogens or pollution of the water supply can cause infections as a result.

1.18 SEWAGE CHARACTERISTICS

1.18.1 Types of Sewage

There are three types of wastewater, or sewage: domestic sewage, industrial sewage, and storm sewage. Domestic sewage carries used water from houses and apartments; it is also called sanitary sewage. Industrial sewage is used water from manufacturing or chemical processes. Storm sewage, or storm water, is runoff from precipitation that is collected in a system of pipes or open channels.

Domestic sewage is slightly more than 99.9 percent water by weight. The rest, less than 0.1 percent, contains a wide variety of dissolved and suspended impurities. Although amounting to a very small fraction of the sewage by weight, the nature of these impurities and the large volumes of sewage in which they are carried make disposal of domestic wastewater a significant technical problem. The principal impurities are putrescible organic materials and plant nutrients, but domestic sewage is also very likely to contain disease-causing microbes. Industrial wastewater usually contains specific and readily identifiable chemical compounds, depending on the nature of the industrial process. Storm sewage carries organic materials, suspended and dissolved solids, and other substances picked up as it travels over the ground.

1.19 PRINCIPAL POLLUTANTS

1.19.1 Organic Material

The amount of putrescible organic material in sewage is indicated by the biochemical oxygen demand, or BOD; the more organic material there is in the sewage, the higher the BOD, which is the amount of oxygen required by microorganisms to decompose the organic substances in sewage. It is among the most important parameters for the design and operation of sewage treatment plants. Industrial sewage may have BOD levels many times that of domestic sewage. The BOD of storm sewage is of particular concern when it is mixed with domestic sewage in combined sewerage systems.

Dissolved oxygen is an important water quality factor for lakes and rivers. The higher the concentration of dissolved oxygen, the better the water quality. When sewage enters a lake or stream, decomposition of the organic materials begins. Oxygen is consumed as microorganisms use it in their metabolism. This can quickly deplete the available oxygen in the water. When the dissolved oxygen levels drop too low, trout and other aquatic species soon perish. In fact, if the oxygen level drops to zero, the water will become septic. Decomposition of organic compounds without oxygen causes the undesirable odours usually associated with septic or putrid conditions.

1.19.2 Suspended Solids

Another important characteristic of sewage is suspended solids. The volume of sludge produced in a treatment plant is directly related to the total suspended solids present in the sewage. Industrial and storm sewage may contain higher concentrations of suspended solids than domestic sewage. The extent to which a treatment plant removes suspended solids, as well as BOD, determines the efficiency of the treatment process.

1.19.3 Plant Nutrients

Domestic sewage contains compounds of nitrogen and phosphorus, two elements that are basic nutrients essential for the growth of plants. In lakes, excessive amounts of nitrates and phosphates can cause the rapid growth of algae. Algal blooms, often caused by sewage discharges, accelerate the natural aging of lakes in a process called eutrophication.

1.19.4 Microbes

Scientists have developed a unique, easy, and effective method for testing for the presence of the COVID-19 virus in the wastewater flowing through municipal sewer systems.

Domestic sewage contains many millions of microorganisms per gallon. Most are coliform bacteria from the human intestinal tract, and domestic sewage is also likely to carry other microbes. Coliforms are used as indicators of sewage pollution. A high coliform count usually indicates recent sewage pollution.

1.20 WHY TREAT WASTEWATER?

It's a matter of caring for our environment and for our own health. There are a lot of good reasons why keeping our water clean is an important priority:

Fisheries: Clean water is critical to plants and animals that live in water. This is important to the fishing industry, sport fishing enthusiasts, and future generations.

Wildlife Habitats: Our rivers and ocean waters teem with life that depends on shoreline, beaches and marshes. They are critical habitats for hundreds of species of fish and other aquatic life. Migratory water birds use the areas for resting and feeding.

Recreation and Quality of Life: Water is a great playground for us all. The scenic and recreational values of our waters are reasons many people choose to live where they do. Visitors are drawn to water activities such as swimming, fishing, boating and picnicking.

Health Concerns: If it is not properly cleaned, water can carry disease. Since we live, work and play so close to water, harmful bacteria have to be removed to make water safe

1.21 EFFECTS OF WASTEWATER POLLUTANTS

Epic September 2009 flooding around Atlanta, Georgia. An overflowing sewer on Riverside Road, Roswell, Georgia. Likely this is a storm sewer, designed to carry stormwater runoff off of streets, that cannot handle the volume of runoff. In older sections of Atlanta there are combined sewer systems that are sewers that are designed to collect rainwater runoff, domestic sewage, and industrial wastewater in the same pipe. These overflows, called combined sewer overflows (CSOs) contain not only stormwater but also untreated human and industrial waste, toxic materials, and debris. They are a major water pollution concern for the approximately 772 cities in the U.S. that have combined sewer systems (EPA). The City of Atlanta is spending about $3 billion dollars to put in separate storm and waste systems in the metro Atlanta area.

1.21.1 Alan Cressler, USGS

If wastewater is not properly treated, then the environment and human health can be negatively impacted. These impacts can include harm to fish and wildlife populations, oxygen depletion, beach closures and other restrictions on recreational water use, restrictions on fish and shellfish harvesting and contamination of drinking water. Environment Canada provides some examples of pollutants that can be found in wastewater and the potentially harmful effects these substances can have on ecosystems and human health:

- Decaying organic matter and debris can use up the dissolved oxygen in a lake so fish and other aquatic biota cannot survive;

- Excessive nutrients, such as phosphorus and nitrogen (including ammonia), can cause eutrophication, or over-fertilization of receiving waters, which can be toxic to aquatic organisms, promote excessive plant growth, reduce available oxygen, harm spawning grounds, alter habitat and lead to a decline in certain species;
- Chlorine compounds and inorganic chloramines can be toxic to aquatic invertebrates, algae and fish;
- Bacteria, viruses and disease-causing pathogens can pollute beaches and contaminate shellfish populations, leading to restrictions on human recreation, drinking water consumption and shellfish consumption;
- Metals, such as mercury, lead, cadmium, chromium and arsenic can have acute and chronic toxic effects on species.
- Other substances such as some pharmaceutical and personal care products, primarily entering the environment in wastewater effluents, may also pose threats to human health, aquatic life and wildlife.

2

WASTEWATER TREATMENT

2.1 OBJECTIVES

After studying this chapter you will be able to understand:

- Sewarage systems
- Wastewater treatment and disposal
- Activated sludge
- Removal of plant nutrients
- Wastewater reuse

2.2 INTRODUCTION

The major aim of wastewater treatment is to remove as much of the suspended solids as possible before the remaining water, called effluent, is discharged back to the environment. As solid material decays, it uses up oxygen, which is needed by the plants and animals living in the water.

"Primary treatment" removes about 60 percent of suspended solids from wastewater. This treatment also involves aerating (stirring up) the wastewater, to put oxygen back in. Secondary treatment removes more than 90 percent of suspended solids.

2.3 SEWERAGE SYSTEMS

A sewerage system, or wastewater collection system, is a network of pipes, pumping stations, and appurtenances that convey sewage from its points of origin to a point of treatment and disposal.

2.3.1 Combined Systems

Systems that carry a mixture of both domestic sewage and storm sewage are called combined sewers. Combined sewers typically consist of large-diameter pipes or tunnels, because of the large volumes of storm water that must be carried during wet-weather periods. They are very common in older cities but are no longer designed and built as part of new sewerage facilities. Because wastewater treatment plants cannot handle large volumes of storm water, sewage must bypass the treatment plants during wet weather and be discharged directly into the receiving water. These combined sewer overflows, containing untreated domestic sewage, cause recurring water pollution problems and are very troublesome sources of pollution.

In some large cities the combined sewer overflow problem has been reduced by diverting the first flush of combined sewage into a large basin or underground tunnel. After temporary storage, it can be treated by settling and disinfection before being discharged into a receiving body of water, or it can be treated in a nearby wastewater treatment plant at a rate that will not overload the facility. Another method for controlling combined sewage involves the use of swirl concentrators. These direct sewage through cylindrically shaped devices that create a vortex, or whirlpool, effect. The vortex helps concentrate impurities in a much smaller volume of water for treatment.

2.3.2 Separate Systems

New wastewater collection facilities are designed as separate systems, carrying either domestic sewage or storm sewage but not both. Storm sewers usually carry surface runoff to a point of disposal in a stream or river. Small detention basins may be built as part of the system, storing storm water temporarily and reducing the magnitude of the peak flow rate. Sanitary sewers, on the other hand, carry domestic wastewater to a sewage treatment plant. Pretreated industrial wastewater may be allowed into municipal sanitary sewerage systems, but storm water is excluded.

Storm sewers are usually built with sections of reinforced concrete pipe. Corrugated metal pipes may be used in some cases. Storm water inlets or catch basins are located at suitable intervals in a street right-of-way or in easements across private property. The pipelines are usually located to allow downhill gravity flow to a nearby stream or to a detention basin. Storm water pumping stations are avoided, if possible, because of the very large pump capacities that would be needed to handle the intermittent flows.

A sanitary sewerage system includes laterals, submains, and interceptors. Except for individual house connections, laterals are the smallest sewers in the network. They usually are not less than 200 mm (8 inches) in diameter and carry sewage by gravity into larger submains, or collector sewers. The collector sewers tie in to a main interceptor, or trunk line, which carries the sewage to a treatment plant. Interceptors are usually built with precast sections of reinforced concrete pipe, up to 5 metres (15 feet) in diameter. Other materials used for sanitary sewers include vitrified clay, asbestos cement, plastic, steel, or ductile iron. The use of plastic for laterals is increasing because of its lightness and ease of installation. Iron and steel pipes are used for force mains or in pumping stations. Force mains are pipelines that carry sewage under pressure when it must be pumped.

2.3.3 Alternative Systems

Sometimes the cost of conventional gravity sewers can be prohibitively high because of low population densities or site conditions such as a high water table or bedrock. Three alternative wastewater collection systems that may be used under these circumstances include small-diameter gravity sewers, pressure sewers, and vacuum sewers.

In small-diameter gravity systems, septic tanks are first used to remove settleable and floating solids from the wastewater from each house before it flows into a network of collector mains (typically 100 mm, or 4 inches, in diameter); these systems are most suitable for small rural communities. Because they do not carry grease, grit and sewage solids, the pipes can be of smaller diameter and placed at reduced slopes or gradients to minimize trench excavation costs. Pressure sewers are best used in flat areas or where expensive rock excavation would be required. Grinder pumps discharge wastewater from each home into the main pressure sewer, which can follow the slope of the ground. In a vacuum sewerage system, sewage from one or more buildings flows by gravity into a sump or tank from which it is pulled out by vacuum pumps located at a central vacuum station and then flows into a collection tank. From the vacuum collection tank the sewage is pumped to a treatment plant.

2.4 PUMPS

Pumping stations are built when sewage must be raised from a low point to a point of higher elevation or where the topography prevents downhill gravity flow. Special nonclogging pumps are available to handle raw sewage. They are installed in structures called lift stations. There are two basic types of lift stations: dry well and wet well. A wet-well installation has only one chamber or tank to receive and hold the sewage until it is pumped out. Specially designed submersible pumps and motors can be located at the bottom of the chamber, completely below the water level. Dry-well installations have two separate chambers, one to receive the wastewater and one to enclose and protect the pumps and controls. The protective dry chamber allows easy access for inspection and maintenance. All sewage lift stations, whether of the wet-well or dry-well type, should include at least two pumps. One pump can operate while the other is removed for repair.

2.4.1 Flow Rates

There is a wide variation in sewage flow rates over the course of a day. A sewerage system must accommodate this variation. In most cities domestic sewage flow rates are highest in the morning and evening hours. They are lowest during the middle of the night. Flow quantities depend upon population density, water consumption, and the extent of commercial or industrial activity in the community. The average sewage flow rate is usually about the same as the average water use in the community. In a lateral sewer, short-term peak flow rates can be roughly four times the average flow rate. In a trunk sewer, peak flow rates may be two-and-a-half times the average.

Although sewage flows depend upon residential, commercial, and industrial connections, sewage flow rates potentially can become higher as a result of inflows and infiltration (I&I) into the sanitary sewer system. Inflows correspond to storm water entering sewers from inappropriate connections, such as roof drains, storm drains, downspouts and sump pumps. High amounts of rainwater runoff can reach the sewer system during precipitation and stormflow events or during seasonal spring flooding of rivers inundated with melting ice. Infiltration refers to the groundwater entering sewers via defective or broken pipes. In both these cases, downstream utilities and treatment plants may experience flows higher than anticipated and can become hydraulically overloaded. During such overloads, utilities may ask residents connected to the system to refrain from using dishwashers and washing machines and may even limit toilet flushing and the use of showers in an attempt to lessen the strain. Such I&I issues can be especially severe in old and aging water infrastructures.

2.5 WASTEWATER TREATMENT AND DISPOSAL

The size and capacity of wastewater treatment systems are determined by the estimated volume of sewage generated from residences, businesses, and industries connected to sewer systems as well as the anticipated Inflows and Infiltration (I&I). The selection of specific on-lot, clustered, or centralized treatment plant configurations depends upon factors such as the number of customers being served, the geographical scenario, site constraints, sewer connections, average and peak flows, influent wastewater characteristics, regulatory effluent limits, technological feasibility, energy consumption, and the operations and maintenance costs involved.

The predominant method of wastewater disposal in large cities and towns is discharge into a body of surface water. Suburban and rural areas rely more on subsurface disposal. In either case, wastewater must be purified or treated to some degree in order to protect both public health and water quality. Suspended particulates and biodegradable organics must be removed to varying extents. Pathogenic bacteria must be destroyed. It may also be necessary to remove nitrates and phosphates (plant nutrients) and to neutralize or remove industrial wastes and toxic chemicals.

The degree to which wastewater must be treated varies, depending on local environmental conditions and governmental standards. Two pertinent types of standards are stream standards and effluent standards. Stream standards, designed to prevent the deterioration of existing water quality, set limits on the amounts of specific pollutants allowed in streams, rivers, and lakes. The limits depend on a classification of the "maximum beneficial use" of the water. Water quality parameters that are regulated by stream standards include dissolved oxygen, coliforms, turbidity, acidity, and toxic substances. Effluent standards, on the other hand, pertain directly to the quality of the treated wastewater discharged from a sewage treatment plant. The factors controlled under these standards usually include Biochemical Oxygen Demand (BOD), suspended solids, acidity, and coliforms.

There are three levels of wastewater treatment: primary, secondary, and tertiary (or advanced). Primary treatment removes about 60 percent of total suspended solids and about 35 percent of BOD; dissolved impurities are not removed. It is usually used as a first step before secondary treatment. Secondary treatment removes more than 85 percent of both suspended solids and BOD. A minimum level of secondary treatment is usually required in the United States and other developed countries. When more than 85 percent of total solids and BOD must be removed, or when dissolved nitrate and phosphate levels must be reduced, tertiary treatment methods are used. Tertiary processes can remove more than 99 percent of all the impurities from sewage, producing an effluent of almost drinking-water quality. Tertiary treatment can be very expensive, often doubling the cost of secondary treatment. It is used only under special circumstances.

For all levels of wastewater treatment, the last step prior to discharge of the sewage effluent into a body of surface water is disinfection, which destroys any remaining pathogens in the effluent and protects public health. Disinfection is usually accomplished by mixing the effluent with chlorine gas or with liquid solutions of hypochlorite chemicals in a contact tank for at least 15 minutes. Because chlorine residuals in the effluent may have adverse effects on aquatic life, an additional chemical may be added to dechlorinate the effluent. Ultraviolet radiation, which can disinfect without leaving any residual in the effluent, is becoming more competitive with chlorine as a wastewater disinfectant.

2.5.1 Primary Treatment

Primary treatment removes material that will either float or readily settle out by gravity. It includes the physical processes of screening, comminution, grit removal, and sedimentation. Screens are made of long, closely spaced, narrow metal bars. They block floating debris such as wood, rags, and other bulky objects that could clog pipes or pumps. In modern plants the screens are cleaned mechanically, and the material is promptly disposed of by burial on the plant grounds. A comminutor may be used to grind and shred debris that passes through the screens. The shredded material is removed later by sedimentation or flotation processes.

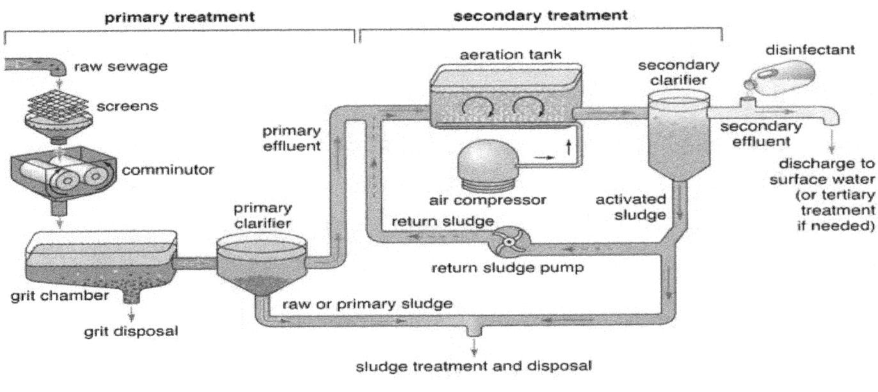

Fig. 1. Primary and secondary treatment of sewage, using the activated sludge process.

Grit chambers are long narrow tanks that are designed to slow down the flow so that solids such as sand, coffee grounds, and eggshells will settle out of the water. Grit causes excessive wear and tear on pumps and other plant equipment. Its removal is particularly important in cities with combined sewer systems, which carry a good deal of silt, sand, and gravel that wash off streets or land during a storm.

Suspended solids that pass through screens and grit chambers are removed from the sewage in sedimentation tanks. These tanks, also called primary clarifiers, provide about two hours of detention time for gravity settling to take place. As the sewage flows through them slowly, the solids gradually sink to the bottom. The settled solids—known as raw or primary sludge—are moved along the tank bottom by mechanical scrapers. Sludge is collected in a hopper, where it is pumped out for removal. Mechanical surface-skimming devices remove grease and other floating materials.

2.5.2 Secondary Treatment

Secondary treatment removes the soluble organic matter that escapes primary treatment. It also removes more of the suspended solids. Removal is usually accomplished by biological processes in which microbes consume the organic impurities as food, converting them into carbon dioxide, water, and energy for their own growth and reproduction. The sewage treatment plant provides a suitable environment, albeit of steel and concrete, for this natural biological process. Removal of soluble organic matter at the treatment plant helps to protect the dissolved oxygen balance of a receiving stream, river, or lake.

There are three basic biological treatment methods: the trickling filter, the activated sludge process, and the oxidation pond. A fourth, less common method is the rotating biological contacter.

Trickling Filter

A trickling filter is simply a tank filled with a deep bed of stones. Settled sewage is sprayed continuously over the top of the stones and trickles to the bottom, where it is collected for further treatment. As the wastewater trickles down, bacteria gather and multiply on the stones. The steady flow of sewage over these growths allows the microbes to absorb the dissolved organics, thus lowering the Biochemical Oxygen Demand (BOD) of the sewage. Air circulating upward through the spaces among the stones provides sufficient oxygen for the metabolic processes.

Settling tanks, called secondary clarifiers, follow the trickling filters. These clarifiers remove microbes that are washed off the rocks by the flow of wastewater. Two or more trickling filters may be connected in series, and sewage can be recirculated in order to increase treatment efficiencies.

2.5.3 Tertiary Treatment

When the intended receiving water is very vulnerable to the effects of pollution, secondary effluent may be treated further by several tertiary processes.

2.6 ACTIVATED SLUDGE

The activated sludge treatment system consists of an aeration tank followed by a secondary clarifier. Settled sewage, mixed with fresh sludge that is recirculated from the secondary clarifier, is introduced into the aeration tank. Compressed air is then injected into the mixture through porous diffusers located at the bottom of the tank. As it bubbles to the surface, the diffused air provides oxygen and a rapid mixing action. Air can also be added by the churning action of mechanical propeller-like mixers located at the tank surface.

Under such oxygenated conditions, microorganisms thrive, forming an active, healthy suspension of biological solids—mostly bacteria—called activated sludge. About six hours of detention is provided in the aeration tank. This gives the microbes enough time to absorb dissolved organics from the sewage, reducing the BOD. The mixture then flows from the aeration tank into the secondary clarifier, where activated sludge settles out by gravity. Clear water is skimmed from the surface of the clarifier, disinfected, and discharged as secondary effluent. The sludge is pumped out from a hopper at the bottom of the tank. About 30 percent of the sludge is recirculated back into the aeration tank, where it is mixed with the primary effluent. This recirculation is a key feature of the activated sludge process. The recycled microbes are well acclimated to the sewage environment and readily metabolize the organic materials in the primary effluent. The remaining 70 percent of the secondary sludge must be treated and disposed of in an acceptable manner.

Variations of the activated sludge process include extended aeration, contact stabilization, and high-purity oxygen aeration. Extended aeration and contact stabilization systems omit the primary settling step. They are efficient for treating small sewage flows from motels, schools, and other relatively isolated wastewater sources. Both of these treatments are usually provided in prefabricated steel tanks called package plants. Oxygen aeration systems mix pure oxygen with activated sludge. A richer concentration of oxygen allows the aeration time to be shortened from six to two hours, reducing the required tank volume.

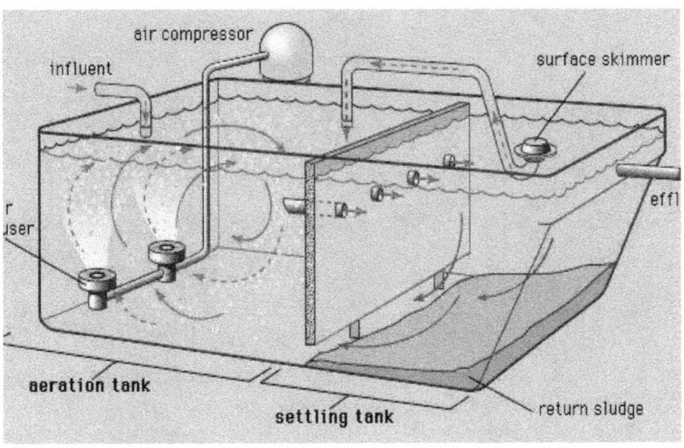

Fig. 2. Aeration treatment Schematic diagram of a prefabricated package plant for the aeration treatment of small sewage flows.

2.7 OXIDATION POND

Oxidation ponds, also called lagoons or stabilization ponds, are large, shallow ponds designed to treat wastewater through the interaction of sunlight, bacteria, and algae. Algae grow using energy from the sun and carbon dioxide and inorganic compounds released by bacteria in water. During the process of photosynthesis, the algae release oxygen needed by aerobic bacteria. Mechanical aerators are sometimes installed to supply yet more oxygen, thereby reducing the required size of the pond. Sludge deposits in the pond must eventually be removed by dredging. Algae remaining in the pond effluent can be removed by filtration or by a combination of chemical treatment and settling.

2.7.1 Rotating Biological Contactor

In this treatment system a series of large plastic disks mounted on a horizontal shaft are partially submerged in primary effluent. As the shaft rotates, the disks are exposed alternately to air and wastewater, allowing a layer of bacteria to grow on the disks and to metabolize the organics in the wastewater.

2.7.2 Effluent Polishing

For the removal of additional suspended solids and BOD from secondary effluent, effluent polishing is an effective treatment. It is most often accomplished using granular media filters, much like the filters used to purify drinking water. Polishing filters are usually built as prefabricated units, with tanks placed directly above the filters for storing backwash water. Effluent polishing of wastewater may also be achieved using microstrainers of the type used in treating municipal water supplies.

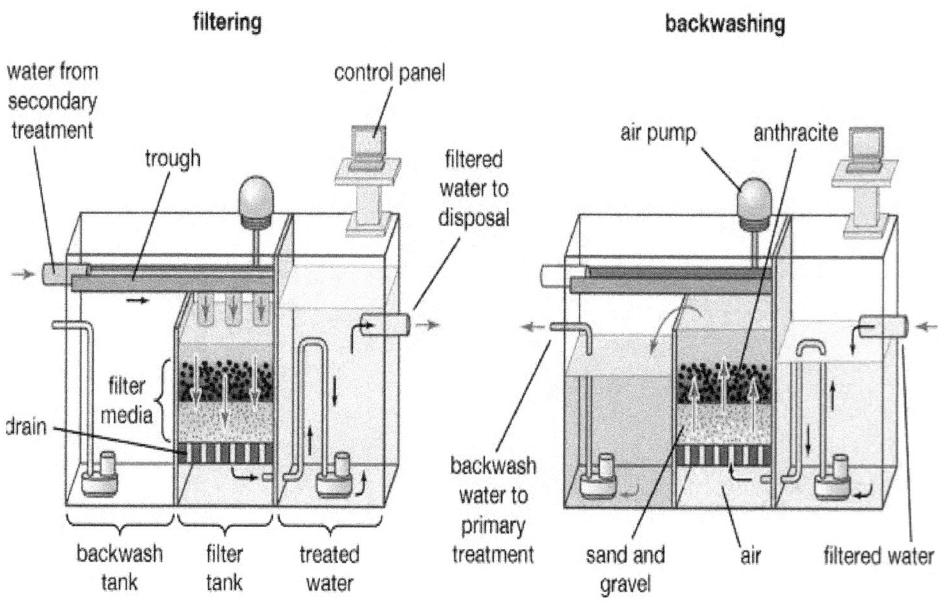

Fig. 3. Tertiary treatment of wastewater.

During the filtering step, wastewater from secondary treatment, still containing suspended solids, pours from a trough and percolates through a filter bed made of porous media such as sand, gravel, and anthracite. The filtered water is then piped away for disposal. (Right) In the backwashing step, entrained solids are periodically flushed from the filter media by pumping filtered water back through the assembly. The backwash water, carrying suspended solids, is returned to the beginning of the wastewater treatment process.

2.8 REMOVAL OF PLANT NUTRIENTS

When treatment standards require the removal of plant nutrients from the sewage, it is often done as a tertiary step. Phosphorus in wastewater is usually present in the form of organic compounds and phosphates that can easily be removed by chemical precipitation. This process, however, increases the volume and weight of sludge. Nitrogen, another important plant nutrient, is present in sewage in the form of ammonia and nitrates. Ammonia is toxic to fish, and it also exerts an oxygen demand in receiving waters as it is converted to nitrates. Nitrates, like phosphates, promote the growth of algae and the eutrophication of lakes. A method called nitrification-denitrification can be used to remove the nitrates. It is a two-step biological process in which ammonia nitrogen is first converted into nitrates by microorganisms. The nitrates are further metabolized by another species of bacteria, forming nitrogen gas that escapes into the air. This process requires the construction of more aeration and settling tanks and significantly increases the cost of treatment.

A physicochemical process called ammonia stripping may be used to remove ammonia from sewage. Chemicals are added to convert ammonium ions into ammonia gas. The sewage is then cascaded down through a tower, allowing the gas to come out of solution and escape into the air. Stripping is less expensive than nitrification-denitrification, but it does not work very efficiently in cold weather.

2.8.1 Land Treatment

In some locations, secondary effluent can be applied directly to the ground and a polished effluent obtained by natural processes as the wastewater flows over vegetation and percolates through the soil. There are three types of land treatment: slow-rate, rapid infiltration, and overland flow.

In the slow-rate, or irrigation, method, effluent is applied onto the land by ridge-and-furrow spreading (in ditches) or by sprinkler systems. Most of the water and nutrients are absorbed by the roots of growing vegetation. In the rapid infiltration method, the wastewater is stored in large ponds called recharge basins. Most of it percolates to the groundwater, and very little is absorbed by vegetation. For this method to work, soils must be highly permeable. In overland flow, wastewater is sprayed onto an inclined vegetated terrace and slowly flows to a collection ditch. Purification is achieved by physical, chemical, and biological processes, and the collected water is usually discharged into a nearby stream.

Land treatment of sewage can provide moisture and nutrients for the growth of vegetation, such as corn or grain for animal feed. It also can recharge, or replenish, groundwater aquifers. Land treatment, in effect, allows sewage to be recycled for beneficial use. Large land areas are required, however, and the feasibility of this kind of treatment may be limited further by soil texture and climate.

2.9 CLUSTERED WASTEWATER TREATMENT SYSTEMS

In certain instances when it is not feasible to connect residences or units to public sewer systems, communities may opt for a clustered wastewater treatment system. Such facilities are smaller versions of centralized treatment plants and serve only a limited number of connections. The technologies used for clustered wastewater treatment may be the same as those used for centralized systems or for individual on-site systems, depending upon the specific applications and degree of treatment required. Upon treatment, effluent from clustered wastewater systems can be discharged via surface or subsurface disposal methods.

2.9.1 On-Site Septic Tanks and Leaching Fields

In sparsely populated suburban or rural areas, it is usually not economical to build sewage collection systems and a centrally located treatment plant. Instead, a separate treatment and disposal system is provided for each home. On-site systems provide effective, low-cost, long-term solutions for wastewater disposal as long as they are properly designed, installed, and maintained. In the United States, about one-third of private homes make use of an on-site subsurface disposal system.

The most common type of on-site system includes a buried, watertight septic tank and a subsurface absorption field (also called a drain field or leaching field). The septic tank serves as a primary sedimentation and sludge storage chamber, removing most of the settleable and floating material from the influent wastewater. Although the sludge decomposes anaerobically, it eventually accumulates at the tank bottom and must be pumped out periodically (every two to four years).

Floating solids and grease are trapped by a baffle at the tank outlet, and settled sewage flows out into the absorption field, through which it percolates downward into the ground. As it flows slowly through layers of soil, the settled wastewater is further treated and purified by both physical and biological processes before it reaches the water table.

An absorption field includes several perforated pipelines placed in long, shallow trenches filled with gravel. The pipes distribute the effluent over a sizable area as it seeps through the gravel and into the underlying layers of soil. If the disposal site is too small for a conventional leaching field, deeper seepage pits may be used instead of shallow trenches; seepage pits require less land area than leaching fields. Both leaching field trenches and seepage pits must be placed above seasonally high groundwater levels.

Fig. 4. A septic tank before installation.

For subsurface on-site wastewater disposal to succeed, the permeability, or hydraulic conductivity, of the soil must be within an acceptable range. If it is too low, the effluent will not be able to flow effectively through the soil, and it may seep out onto the surface of the absorption field, thereby endangering public health. If permeability is too high, there may not be sufficient purification before the effluent reaches the water table, thereby contaminating the groundwater. The capacity of the ground to absorb settled wastewater depends largely on the texture of the soil (i.e., relative amounts of gravel, sand, silt, and clay). Permeability can be evaluated by direct observation of the soil in excavated test pits and also by conducting a percolation test, or "per test." The perc test measures the rate at which water seeps into the soil in small test holes dug on the disposal site. The measured perc rate can be used to determine the total required area of the absorption field or the number of seepage pits.

Where unfavourable site or soil conditions prohibit the use of both absorption fields and seepage pits, mound systems may be utilized for on-site sewage disposal. A mound is an absorption field built above the natural ground surface in order to provide suitable material for percolation and to separate the drain field from the water table. Septic tank effluent is intermittently pumped from a chamber and applied to the mound. Other alternative on-site disposal methods include use of intermittent sand filters or of small, prefabricated aerobic treatment units. Disinfection (usually by chlorination) of the effluent from these systems is required when the effluent is discharged into a nearby stream.

2.10 WASTEWATER REUSE

Wastewater can be a valuable resource in cities or towns where population is growing and water supplies are limited. In addition to easing the strain on limited freshwater supplies, the reuse of wastewater can improve the quality of streams and lakes by reducing the effluent discharges that they receive. Wastewater may be reclaimed and reused for crop and landscape irrigation, groundwater recharge, or recreational purposes. Reclamation for drinking is technically possible, but this reuse faces significant public resistance.

There are two types of wastewater reuse: direct and indirect. In direct reuse, treated wastewater is piped into some type of water system without first being diluted in a natural stream or lake or in groundwater. One example is the irrigation of a golf course with effluent from a municipal wastewater treatment plant. Indirect reuse involves the mixing of reclaimed wastewater with another body of water before reuse. In effect, any community that uses a surface water supply downstream from the treatment plant discharge pipe of another community is indirectly reusing wastewater. Indirect reuse is also accomplished by discharging reclaimed wastewater into a groundwater aquifer and later withdrawing the water for use. Discharge into an aquifer (called artificial recharge) is done by either deep-well injection or shallow surface spreading.

Quality and treatment requirements for reclaimed wastewater become more stringent as the chances for direct human contact and ingestion increase. The impurities that must be removed depend on the intended use of the water. For example, removal of phosphates or nitrates is not necessary if the intended use is landscape irrigation. If direct reuse as a potable supply is intended, tertiary treatment with multiple barriers against contaminants is required. This may include secondary treatment followed by granular media filtration, ultraviolet radiation, granular activated carbon adsorption, reverse osmosis, air stripping, ozonation, and chlorination.

The use of gray-water recycling systems in new commercial buildings offers a method of saving water and reducing total sewage volumes. These systems filter and chlorinate drainage from tubs and sinks and reuse the water for nonpotable purposes (e.g., flushing toilets and urinals). Recycled water can be marked with a blue dye to ensure that it is not used for potable purposes.

2.11 SLUDGE TREATMENT AND DISPOSAL

The residue that accumulates in sewage treatment plants is called sludge (or biosolids). Sewage sludge is the solid, semisolid, or slurry residual material that is produced as a by-product of wastewater treatment processes. This residue is commonly classified as primary and secondary sludge. Primary sludge is generated from chemical precipitation, sedimentation, and other primary processes, whereas secondary sludge is the activated waste biomass resulting from biological treatments. Some sewage plants also receive septage or septic tank solids from household on-site wastewater treatment systems. Quite often the sludges are combined together for further treatment and disposal.

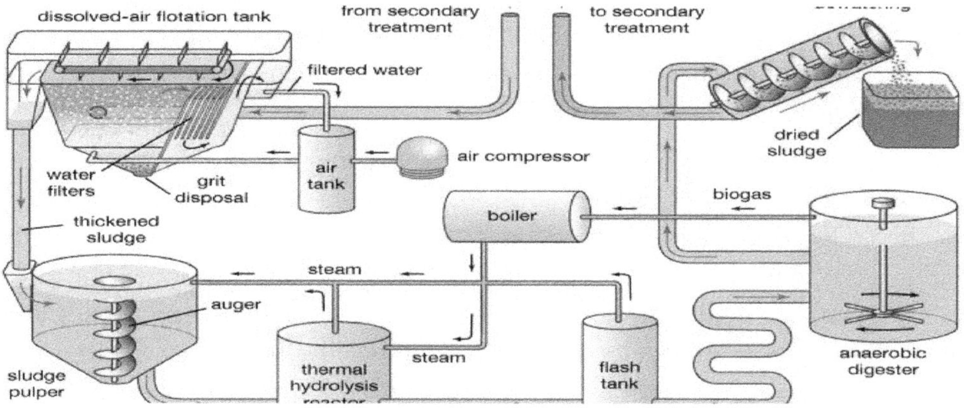

Fig. 5. Sewage sludge treatment.

Mixed sludge received from secondary wastewater treatment is passed through a dissolved-air flotation tank, where solids rise to the surface and are skimmed off. The thickened sludge is pulped with steam, then passed to thermal hydrolysis, where large molecules such as proteins and lipids are broken down under heat and pressure. The hydrolyzed sludge is passed through a flash tank, where a sudden drop in pressure causes cells to burst, and then to anaerobic digestion, where bacteria convert dissolved organic matter to biogas (which can be used to fuel the treatment process). Digested sludge is passed through a dewatering step; the dried solids are disposed of, and the water is sent back to secondary treatment.

Treatment and disposal of sewage sludge are major factors in the design and operation of all wastewater treatment plants. Two basic goals of treating sludge before final disposal are to reduce its volume and to stabilize the organic materials. Stabilized sludge does not have an offensive odour and can be handled without causing a nuisance or health hazard. Smaller sludge volume reduces the costs of pumping and storage.

2.12 TREATMENT METHODS

Treatment of sewage sludge may include a combination of thickening, digestion, and dewatering processes.

2.12.1 Thickening

Thickening is usually the first step in sludge treatment because it is impractical to handle thin sludge, a slurry of solids suspended in water. Thickening is usually accomplished in a tank called a gravity thickener. A thickener can reduce the total volume of sludge to less than half the original volume. An alternative to gravity thickening is dissolved-air flotation. In this method, air bubbles carry the solids to the surface, where a layer of thickened sludge forms.

2.12.2 Digestion

Sludge digestion is a biological process in which organic solids are decomposed into stable substances. Digestion reduces the total mass of solids, destroys pathogens, and makes it easier to dewater or dry the sludge. Digested sludge is inoffensive, having the appearance and characteristics of a rich potting soil.

Most large sewage treatment plants use a two-stage digestion system in which organics are metabolized by bacteria anaerobically (in the absence of oxygen). In the first stage, the sludge, thickened to a Dry Solids (DS) content of about 5 percent, is heated and mixed in a closed tank for several days. Acid-forming bacteria hydrolyze large molecules such as proteins and lipids, breaking them into smaller water-soluble molecules, and then ferment those smaller molecules into various fatty acids. The sludge then flows into a second tank, where the dissolved matter is converted by other bacteria into biogas, a mixture of carbon dioxide and methane. Methane is combustible and is used as a fuel to heat the first digestion tank as well as to generate electricity for the plant.

Anaerobic digestion is very sensitive to temperature, acidity, and other factors. It requires careful monitoring and control. In some cases, the sludge is inoculated with extra hydrolytic enzymes at the beginning of the first digestion stage in order to supplement the action of the bacteria. It has been found that this enzymatic treatment can destroy more unwanted pathogens in the sludge and also can result in the generation of more biogas in the second stage of digestion.

Another enhancement of the traditional two-stage anaerobic digestion process is thermal hydrolysis, or the breaking down of the large molecules by heat. This is done in a separate step before digestion. In a typical case, the process begins with a sludge that has been dewatered to a DS content of some 15 percent. The sludge is mixed with steam in a pulper, and this hot homogenized mixture is fed to a reactor, where it is held under pressure at approximately 165 °C (about 330 °F) for about 30 minutes. At that point, with the hydrolytic reactions complete, some of the steam is bled off (to be fed to the pulper), and the sludge, still under some pressure, is released suddenly into a "flash tank," where the sudded drop in pressure bursts the cell walls of much of the solid matter. The hydrolyzed sludge is cooled, diluted slightly with water, and then sent directly to the second stage of anaerobic digestion.

Sludge digestion may also take place aerobically—that is, in the presence of oxygen. The sludge is vigorously aerated in an open tank for about 20 days. Methane gas is not formed in this process. Although aerobic systems are easier to operate than anaerobic systems, they usually cost more to operate because of the power needed for aeration. Aerobic digestion is often combined with small extended aeration or contact stabilization systems.

Aerobic and conventional anaerobic digestion convert about half of the organic sludge solids to liquids and gases. Thermal hydrolysis followed by anaerobic digestion can convert some 60 to 70 percent of the solid matter to liquids and gases. Not only is the volume of solids produced smaller than in conventional digestion, but the greater production of biogas can make some wastewater treatment plants self-sufficient in energy.

2.12.3 Dewatering

Digested sewage sludge is usually dewatered before disposal. Dewatered sludge still contains a significant amount of water—often as much as 70 percent—but, even with that moisture content, sludge no longer behaves as a liquid and can be handled as a solid material. Sludge-drying beds provide the simplest method of dewatering. A digested sludge slurry is spread on an open bed of sand and allowed to remain until dry. Drying takes place by a combination of evaporation and gravity drainage through the sand. A piping network built under the sand collects the water, which is pumped back to the head of the plant. After about six weeks of drying, the sludge cake, as it is called, may have a solids content of about 40 percent. It can then be removed from the sand with a pitchfork or a front-end loader. In order to reduce drying time in wet or cold weather, a glass enclosure may be built over the sand beds. Since a good deal of land area is needed for drying beds, this method of dewatering is commonly used in rural or suburban towns rather than in densely populated cities.

Alternatives to sludge-drying beds include the rotary drum vacuum filter, the centrifuge, and the belt filter press. These mechanical systems require less space than do sludge-drying beds, and they offer a greater degree of operational control. However, they usually have to be preceded by a step called sludge conditioning, in which chemicals are added to the liquid sludge to coagulate solids and improve drainability.

Disposal

The final destination of treated sewage sludge usually is the land. Dewatered sludge can be buried underground in a sanitary landfill. It also may be spread on agricultural land in order to make use of its value as a soil conditioner and fertilizer. Since sludge may contain toxic industrial chemicals, it is not spread on land where crops are grown for human consumption.

Where a suitable site for land disposal is not available, as in urban areas, sludge may be incinerated. Incineration completely evaporates the moisture and converts the organic solids into inert ash. The ash must be disposed of, but the reduced volume makes disposal more economical. Air pollution control is a very important consideration when sewage sludge is incinerated. Appropriate air-cleaning devices such as scrubbers and filters must be used.

Dumping sludge in the ocean, once an economical disposal method for many coastal communities, is no longer considered a viable option. It is now prohibited in the United States and many other coastal countries.

2.13 EMERGING TECHNOLOGIES

Experts in the wastewater treatment sector have been working to implement established technologies and to improve environmental rules and regulations to meet water quality goals and human health protection. At the same time, the industry has also been transitioning to prepare for future challenges, such as climate change, changing populations, and aging infrastructure.

2.13.1 Improved Treatment Methods

Many older wastewater treatment facilities require upgrading because of increasingly strict water quality standards, but this is often difficult because of limited space for expansion. In order to allow improvement of treatment efficiencies without requiring more land area, new treatment methods have been developed. These include the membrane bioreactor process, the ballasted floc reactor, and the Integrated Fixed-Film Activated sludge (IFAS) process.

In the membrane bioreactor process, hollow-fibre microfiltration membrane modules are submerged in a single tank in which aeration, secondary clarification, and filtration can occur, thereby providing both secondary and tertiary treatment in a small land area.

In a ballasted floc reactor, the settling rate of suspended solids is increased by using sand and a polymer to help coagulate the suspended solids and form larger masses called flocs. The sand is separated from the sludge in a hydroclone, a relatively simple apparatus into which the water is introduced near the top of a cylinder at a tangent so that heavy materials such as sand are "spun" by centrifugal force toward the outside wall. The sand collects by gravity at the bottom of the hydroclone and is recycled back to the reactor.

Biological aerated filters use a basin with submerged media that serves as both a contact surface for biological treatment and a filter to separate solids from the wastewater. Fine-bubble aeration is applied to facilitate the process, and routine backwashing is used to clean the media. The land area required for a biological aerated filter is only about 15 percent of the area required for a conventional activated sludge system.

2.13.2 Automation

Advanced wastewater purification processes involve biological treatments that are sensitive to processing parameters and to the environment. To ensure stable and reliable operations of physical, chemical, and biological processes, treatment plants quite often need to implement sophisticated technologies involving complex instrumentation and process control systems. Use of online analytical instruments, Programmable Logic Controllers (PLC), Supervisory Control and Data Acquisition (SCADA) systems, Human Machine Interface (HMI), and various process control software allow for the automation and computerization of treatment processes with the provision for remote operations. Such innovations improve system operations significantly, thus minimizing supervision needs.

2.13.3 Environmental Considerations

Natural treatments, energy conservation, and carbon footprint reduction are some of the key considerations for communities facing energy and electricity challenges. Green technologies and the use of renewable energy sources, including solar and wind power, for wastewater treatment are evolving and will help minimize the environmental impacts of human activities. Ecological and economical natural wastewater treatment and disposal systems have already gained importance in many places, especially in smaller communities. These include constructed wetlands, lagoons, stabilization ponds, soil filters, drip irrigation, groundwater recharge, and other similar systems. The simplicity, cost-effectiveness, efficiency, and reliability of these systems have provided potential applications for such environmentally friendly technologies.

Given that wastewater is rich in nutrients and other chemicals, sewage treatment facilities have gained recognition as resource recovery facilities, overcoming their former reputation as mere pollution mitigation entities. Newer technologies and approaches have continued to improve the efficiency by which energy, nutrients, and other chemicals are recovered from treatment plants, helping create a sustainable market and becoming a revenue generation source for wastewater processing facilities.

Concepts such as nutrient trading have also emerged. The intention of such initiatives is to control and meet overall pollution load targets for a given watershed by trading nutrient reduction credits between point and non-point source dischargers. Such programs can help to minimize nutrient pollution effects as well as reduce financial burdens on societies for costly treatment plant upgrades.

2.14 MODERN CHALLENGES

2.14.1 Water Scarcity

Water managers confront new challenges and the need for new technology as water levels decrease due to increasingly frequent and extended droughts. Technologies such as sonar mapping are being used in wells to determine the volume of water that they can hold. For example, the United States Geological Survey and the State of New York worked together to map underground aquifers since the 1980s. Today they have thorough maps of these aquifers to assist in water management.

Desalination plants may be required in the future for those regions hardest hit by water scarcity. Desalination is a process of cleaning water by means of evaporation. Water is evaporated and it passes through membranes. The water is then cooled and condenses allowing it to flow either back into the main water line or out to sea.

2.14.2 Climate Change

Wastewater treatment contributes to global warming in many ways. One of the factors that contributes to global warming is wastewater treatment facilities and their emissions of greenhouse gases. Some of those gases are carbon dioxide, methane, and nitrous oxide. These gases occur because of the decomposition of organic material from the anaerobic bacteria. These bacteria clean the leftover waste. Even if the anaerobic bacteria decomposition produces these gases, the percentage of greenhouse gases that other equipment produce is still greater than the contribution of the anaerobic bacteria. Also, the power usage from those machinery is very high. That is why many facilities are undergoing renovation to use higher levels of anaerobic bacteria compared to other types of equipment.

Impacts of climate change on sanitary engineering vary based on region and the sanitation solutions employed there. In the Arctic, permafrost melting has caused damage to pipes and other infrastructure. In the Northeastern United States, increased precipitation has overwhelmed aging infrastructure not equipped to handle the massive volume of water from heavy precipitation. In the Western United States, prolonged drought has decreased water availability. This has led some wastewater facilities to expand recycled and reclaimed water programs. Climate change has also affected water distribution pipes. Physical stress from climate change-related conditions such as extreme rainfall or drought increases the rate of pipe corrosion, adding to facility cost.

3

STRUCTURE, PROPERTIES AND USES OF WATER

3.1 OBJECTIVES

After studying this chapter you will be able to understand:

- Forms of water.
- Sources of fresh water.
- Increasing water scarcity.
- Deplection of aquifers
- National primary drinking watger regulations.
- Application on agricultural land.

3.2 INTRODUCTION

Water (H_2O) is the most abundant compound on Earth's surface, covering 70 percent of the planet. In nature, water exists in liquid, solid, and gaseous states. It is in dynamic equilibrium between the liquid and gas states atstandard temperature and pressure. At room temperature, it is a tasteless andodorless liquid, nearly colorless with a hint of blue. Many substances dissolve in water and it is commonly referred to as the universal solvent. Because of this, water in nature and in use is rarely pure and some properties may vary from those of the pure substance. However, there are also many compounds that are essentially, if not completely, insoluble in water. Water is the only common substance found naturally in all three common states of matter and it is essential for all life on Earth. Water makes up 55% to 78% of the human body

3.3 FORMS OF WATER

Like many substances, water can take numerous forms that are broadly categorized by phase of matter. The liquid phase is the most common among water's phases (within the

Earth's atmosphere and surface) and is the form that is generally denoted by the word "water." The solid phase of water is known as ice and commonly takes the structure of hard, amalgamatedcrystals, such as ice cubes, or loosely accumulated granular crystals, likesnow. For a list of the many different crystalline and amorphous forms of solid H_2O, see the article ice. The gaseous phase of water is known as water vapor(or steam), and is characterized by water assuming the configuration of a transparent cloud. (Note that visible steam and clouds are, in fact, water in the liquid form as minute droplets suspended in the air.) The fourth state of water, that of a supercritical fluid, is much less common than the other three and only rarely occurs in nature, in extremely uninhabitable conditions. When water achieves a specific critical temperature and a specific critical pressure (647 Kand 22.064 MPa), liquid and gas phase merge to one homogeneous fluid phase, with properties of both gas and liquid. One example of naturally occurring supercritical water is found in the hottest parts of deep waterhydrothermal vents, in which water is heated to the critical temperature by scalding volcanic plumes and achieves the critical pressure because of the crushing weight of the ocean at the extreme depths at which the vents are located. Additionally, anywhere there is volcanic activity below a depth of 2.25 km (1.40 mi) can be expected to have water in the supercritical phase.

Vienna Standard Mean Ocean Water is the current international standard for water isotopes. Naturally occurring water is almost completely composed of the neutron-less hydrogen isotope protium. Only 155 ppm include deuterium (2 H or D), a hydrogen isotope with one neutron, and fewer than 20 parts perquintillion include tritium (3 H or T), which has two.

In keeping with the basic rules of chemical nomenclature, water would have a systematic name of *dihydrogen monoxide*, but this is not among the names published by the International Union of Pure and Applied Chemistryand, rather than being used in a chemical context, the name is almost exclusively used as a humorous way to refer to water.

Heavy water is water with a higher-than-average deuterium content, up to 100%. Chemically, it is similar but not identical to normal water. This is because the nucleus of deuterium is twice as heavy as protium, and this causes noticeable differences in bonding energies. Because water molecules exchange hydrogen atoms with one another, hydrogen deuterium oxide (DOH) is much more common in low-purity heavy water than pure dideuterium monoxide (D_2O). Humans are generally unaware of taste differences, but sometimes report a burning sensation or sweet flavor. Rats, however, are able to avoid heavy water by smell. Toxic to many animals, heavy water is used in the nuclear reactor industry to moderate (slow down)neutrons. Light water reactors are also common, where "light" simply designates normal water.

Light water more specifically refers to Deuterium-Depleted Water (DDW), water in which the deuterium content has been reduced below the standard155 ppm level.

3.3.1 Physics and Chemistry

Water is the chemical substance with chemical formula H_2O: one molecule of water has two hydrogen atoms covalentlybonded to a single oxygen atom. Water is a tasteless, odorless liquid at ambient temperature and pressure, and appears colorless in small quantities, although it has its own intrinsic very light blue hue. Ice also appears colorless, and water vapor is essentially invisible as a gas.

Water is primarily a liquid under standard conditions, which is not predicted from its relationship to other analogous hydrides of the oxygen family in the periodic table, which are gases such as hydrogen sulfide. The elements surrounding oxygen in the periodic table, nitrogen, fluorine, phosphorus, sulfur and chlorine, all combine with hydrogen to produce gases under standard conditions. The reason that water forms a liquid is that oxygen is more electronegative than all of these elements with the exception of fluorine. Oxygen attracts electrons much more strongly than hydrogen, resulting in a net positive charge on the hydrogen atoms, and a net negative charge on the oxygen atom. The presence of a charge on each of these atoms gives each water molecule a net dipole moment. Electrical attraction between water molecules due to this dipole pulls individual molecules closer together, making it more difficult to separate the molecules and therefore raising the boiling point. This attraction is known as hydrogen bonding. The molecules of water are constantly moving in relation to each other, and the hydrogen bonds are continually breaking and reforming at timescales faster than 200 femtoseconds.However, this bond is sufficiently strong to create many of the peculiar properties of water, such as those that make it integral to life. Water can be described as a polar liquid that slightly dissociates disproportionately into the hydronium ion (H_3O+ (aq)) and an associated hydroxide ion (OH" (aq)).

2 H

$2O_{(l)}$ H

3O+

(aq) + OH"

(aq)

The dissociation constant for this dissociation is commonly symbolized as Kw and has a value of about 10 at 25 °C; see "Water (data page)" and "Self-ionization of water" for more information.

Percentage of elements in water by mass: 11.1% hydrogen, 88.9% oxygen.

The self-diffusion coefficient of water is 2.299·10 m·s.

3.3.2 Water Resources

Water resources are sources of water that are useful or potentially useful. Uses of water include agricultural, industrial, household, recreational and environmental activities. The majority of human uses require fresh water.

97 percent of the water on the Earth is salt water and only three percent is fresh water; slightly over two thirds of this is frozen in glaciers and polar ice caps. The remaining unfrozen freshwater is found mainly as groundwater, with only a small fraction present above ground or in the air.

Fresh water is a renewable resource, yet the world's supply ofgroundwater is steadily decreasing, with depletion occurring most prominently in Asia and North America, although it is still unclear how much natural renewal balances this usage, and whether ecosystems are threatened. The framework for allocating water resources to water users (where such a framework exists) is known as water rights.

3.4 SOURCES OF FRESH WATER

3.4.1 Surface Water

Surface water is water in a river, lake or fresh water wetland. Surface water is naturally replenished by precipitation and naturally lost through discharge to theoceans, evaporation, evapotranspiration and sub-surface seepage.

Although the only natural input to any surface water system is precipitation within itswatershed, the total quantity of water in that system at any given time is also dependent on many other factors. These factors include storage capacity in lakes, wetlands and artificial reservoirs, the permeability of the soil beneath these storage bodies, the runoff characteristics of the land in the watershed, the timing of the precipitation and local evaporation rates. All of these factors also affect the proportions of water loss.

Human activities can have a large and sometimes devastating impact on these factors. Humans often increase storage capacity by constructing reservoirs and decrease it by draining wetlands. Humans often increase runoff quantities and velocities by paving areas and channelizing stream flow.

The total quantity of water available at any given time is an important consideration. Some human water users have an intermittent need for water. For example, many farms require large quantities of water in the spring, and no water at all in the winter. To supply such a farm with water, a surface water system may require a large storage capacity to collect water throughout the year and release it in a short period of time. Other users have a continuous need for water, such as a power plant that requires water for cooling. To supply such a power plant with water, a surface water system only needs enough storage capacity to fill in when average stream flow is below the power plant's need.

Nevertheless, over the long term the average rate of precipitation within a watershed is the upper bound for average consumption of natural surface water from that watershed.

Natural surface water can be augmented by importing surface water from another watershed through a canal or pipeline. It can also be artificially augmented from any of the other sources listed here, however in practice the quantities are negligible. Humans can also cause surface water to be "lost" (i.e. become unusable) through pollution.

Brazil is the country estimated to have the largest supply of fresh water in the world, followed by Russia and Canada.

3.4.2 Under River Flow

Throughout the course of a river, the total volume of water transported downstream will often be a combination of the visible free water flow together with a substantial contribution flowing through sub-surface rocks and gravels that underlie the river and its floodplain called the hyporheic zone. For many rivers in large valleys, this unseen component of flow may greatly exceed the visible flow. The hyporheic zone often forms a dynamic interface between surface water and true ground-water receiving water from the ground water when aquifers are fully charged and contributing water to ground-water when ground waters are depleted. This is especially significant in karst areas where pot-holes and underground rivers are common.

3.4.3 Ground Water

Sub-surface water, or groundwater, is fresh water located in the porespace of soil and rocks. It is also water that is flowing within aquifersbelow the water table. Sometimes it is useful to make a distinction between sub-surface water that is closely associated with surface water and deep sub-surface water in an aquifer (sometimes called "fossil water"). Sub-surface water can be thought of in the same terms as surface water: inputs, outputs and storage.

The critical difference is that due to its slow rate of turnover, sub-surface water storage is generally much larger compared to inputs than it is for surface water. This difference makes it easy for humans to use sub-surface water unsustainably for a long time without severe consequences. Nevertheless, over the long term the average rate of seepage above a sub-surface water source is the upper bound for average consumption of water from that source.

The natural input to sub-surface water is seepage from surface water. The natural outputs from sub-surface water aresprings and seepage to the oceans.

If the surface water source is also subject to substantial evaporation, a sub-surface water source may become saline. This situation can occur naturally under endorheic bodies of water, or artificially under irrigated farmland. In coastal areas, human use of a sub-surface water source may cause the direction of seepage to ocean to reverse which can also causesoil salinization.

Humans can also cause sub-surface water to be "lost" (i.e. become unusable) through pollution. Humans can increase the input to a sub-surface water source by building reservoirs or detention ponds.

3.4.4 Frozen Water

Several schemes have been proposed to make use of icebergs as a water source, however to date this has only been done for novelty purposes. Glacier runoff is considered to be surface water.

The Himalayas, which are often called "The Roof of the World", contain some of the most extensive and rough high altitude areas on Earth as well as the greatest area of glaciers and permafrost outside of the poles. Ten of Asia's largest rivers flow from there, and more than a billion people's livelihoods depend on them. To complicate matters, temperatures are rising more rapidly here than the global average. In Nepal the temperature has risen by 0.6 degrees Celsius over the last decade, whereas globally, the Earth has warmed approximately 0.7 degrees Celsius over the last hundred years.

3.4.5 Desalination

Desalination is an artificial process by which saline water (generally sea water) is converted to fresh water. The most common desalination processes are distillation and reverse osmosis. Desalination is currently expensive compared to most alternative sources of water, and only a very small fraction of total human use is satisfied by desalination. It is only economically practical for high-valued uses (such as household and industrial uses) in arid areas. The most extensive use is in the Persian Gulf.

3.5 WATER USES

3.5.1 Agricultural

It is estimated that 70% of worldwide water is used for irrigation, with 15-35% of irrigation withdrawals being unsustainable. It takes around 2,000 - 3,000 litres of water to produce enough food to satisfy one person's daily dietary need. This is a considerable amount, when compared to that required for drinking, which is between two and five litres. To produce food for the now over 7 billion people who inhabit the planet today requires the water that would fill a canal ten metres deep, 100 metres wide and 2100 kilometres long.

3.5.2 Increasing Water Scarcity

Around fifty years ago, the common perception was that water was an infinite resource. At this time, there were fewer than half the current number of people on the planet. People were not as wealthy as today, consumed fewer calories and ate less meat, so less water was needed to produce their food. They required a third of the volume of water we presently take from rivers. Today, the competition for water resources is much more intense. This is because there are now seven billion people on the planet, their consumption of water-thirsty meat and vegetables is rising, and there is increasing competition for water from industry, urbanisation biofuel crops, and water reliant food items. In the future, even more water will be needed to produce food because the Earth's population is forecast to rise to 9 billion by 2050. An additional 2.5 or 3 billion people, choosing to eat fewer cereals and more meat and vegetables could add an additional five million kilometres to the virtual canal mentioned above.

An assessment of water management in agriculture sector was conducted in 2007 by the International Water Management Institute in Sri Lanka to see if the world had sufficient water to provide food for its growing population. It assessed the current availability of water for agriculture on a global scale and mapped out locations suffering from water scarcity. It found that a fifth of the world's people, more than 1.2 billion, live in areas of physical water scarcity, where there is not enough water to meet all demands. One third of the worlds population does not have access to clean drinking water, which is more than 2.3 billion people. A further 1.6 billion people live in areas experiencing economic water scarcity, where the lack of investment in water or insufficient human capacity make it impossible for authorities to satisfy the demand for water. The report found that it would be possible to produce the food required in future, but that continuation of today's food production and environmental trends would lead to crises in many parts of the world. To avoid a global water crisis, farmers will have to strive to increase productivity to meet growing demands for food, while industry and cities find ways to use water more efficiently.

In some areas of the world, irrigation is necessary to grow any crop at all, in other areas it permits more profitable crops to be grown or enhances crop yield. Various irrigation methods involve different trade-offs between crop yield, water consumption and capital cost of equipment and structures. Irrigation methods such as furrow and overhead sprinklerirrigation are usually less expensive but are also typically less efficient, because much of the water evaporates, runs off or drains below the root zone. Other irrigation methods considered to be more efficient include drip or trickle irrigation, surge irrigation, and some types of sprinkler systems where the sprinklers are operated near ground level. These types of systems, while more expensive, usually offer greater potential to minimize runoff, drainage and evaporation. Any system that is improperly managed can be wasteful, all methods have the potential for high efficiencies under suitable conditions, appropriate irrigation timing and management. Some issues that are often insufficiently considered are salinization of sub-surface water and contaminant accumulation leading to water quality declines.

As global populations grow, and as demand for food increases in a world with a fixed water supply, there are efforts under way to learn how to produce more food with less water, through improvements in irrigation methods and technologies, agricultural water management, crop types, and water monitoring. Aquaculture is a small but growing agricultural use of water. Freshwater commercial fisheries may also be considered as agricultural uses of water, but have generally been assigned a lower priority than irrigation.

3.5.3 Industrial

It is estimated that 22% of worldwide water is used in industry. Major industrial users include hydroelectric dams, thermoelectric power plants, which use water for cooling, ore and oil refineries, which use water in chemical processes, and manufacturing plants, which use water as a solvent. Water withdrawal can be very high for certain industries, but consumption is generally much lower than that of agriculture.

Water is used in renewable power generation. Hydroelectric power derives energy from the force of water flowing downhill, driving a turbine connected to a generator. This hydroelectricity is a low-cost, non-polluting, renewable energy source. Significantly, hydroelectric power can also be used for load following unlike most renewable energy sources which are intermittent. Ultimately, the energy in a hydroelectric powerplant is supplied by the sun. Heat from the sun evaporates water, which condenses as rain in higher altitudes and flows downhill. Pumped-storage hydroelectric plants also exist, which use grid electricity to pump water uphill when demand is low, and use the stored water to produce electricity when demand is high.

Hydroelectric power plants generally require the creation of a large artificial lake. Evaporation from this lake is higher than evaporation from a river due to the larger surface area exposed to the elements, resulting in much higher water consumption. The process of driving water through the turbine and tunnels or pipes also briefly removes this water from the natural environment, creating water withdrawal. The impact of this withdrawal on wildlife varies greatly depending on the design of the powerplant.

Pressurized water is used in water blasting and water jet cutters. Also, very high pressure water guns are used for precise cutting. It works very well, is relatively safe, and is not harmful to the environment. It is also used in the cooling of machinery to prevent overheating, or prevent saw blades from overheating. This is generally a very small source of water consumption relative to other uses.

Water is also used in many large scale industrial processes, such as thermoelectric power production, oil refining, fertilizer production and other chemical plant use, and natural gas extraction from shale rock. Discharge of untreated water from industrial uses is pollution. Pollution includes discharged solutes (chemical pollution) and increased water temperature (thermal pollution). Industry requires pure water for many applications and utilizes a variety of purification techniques both in water supply and discharge. Most of this pure water is generated on site, either from natural freshwater or from municipalgrey water. Industrial consumption of water is generally much lower than withdrawal, due to laws requiring industrial grey water to be treated and returned to the environment. Thermoelectric powerplants using cooling towers have high consumption, nearly equal to their withdrawal, as most of the withdrawn water is evaporated as part of the cooling process. The withdrawal, however, is lower than in once-through cooling systems.

3.5.4 Household

It is estimated that 8% of worldwide water use is for household purposes. These includedrinking water, bathing, cooking, sanitation, and gardening. Basic household water requirements have been estimated by Peter Gleick at around 50 liters per person per day, excluding water for gardens. Drinking water is water that is of sufficiently high quality so that it can be consumed or used without risk of immediate or long term harm. Such water is commonly called potable water. In most developed countries, the water supplied to households, commerce and industry is all of drinking water standard even though only a very small proportion is actually consumed or used in food preparation.

3.5.5 Recreation

Recreational water use is usually a very small but growing percentage of total water use. Recreational water use is mostly tied to reservoirs. If a reservoir is kept fuller than it would otherwise be for recreation, then the water retained could be categorized as recreational usage. Release of water from a few reservoirs is also timed to enhance whitewater boating, which also could be considered a recreational usage. Other examples are anglers, water skiers, nature enthusiasts and swimmers.

Recreational usage is usually non-consumptive. Golf courses are often targeted as using excessive amounts of water, especially in drier regions. It is, however, unclear whether recreational irrigation (which would include private gardens) has a noticeable effect on water resources. This is largely due to the unavailability of reliable data. Additionally, many golf courses utilize either primarily or exclusively treated effluent water, which has little impact on potable water availability.

Some governments, including the Californian Government, have labelled golf course usage as agricultural in order to deflect environmentalists' charges of wasting water. However, using the above figures as a basis, the actual statistical effect of this reassignment is close to zero. In Arizona, an organized lobby has been established in the form of the Golf Industry Association, a group focused on educating the public on how golf impacts the environment.

Recreational usage may reduce the availability of water for other users at specific times and places. For example, water retained in a reservoir to allow boating in the late summer is not available to farmers during the spring planting season. Water released for white water rafting may not be available for hydroelectric generation during the time of peak electrical demand.

3.5.6 Environmental

Explicit environment water use is also a very small but growing percentage of total water use. Environmental water may include water stored in impoundments and released for environmental purposes (held environmental water), but more often is water retained in waterways through regulatory limits of abstraction. Environmental water usage includes watering of natural or artificial wetlands, artificial lakes intended to create wildlife habitat, fish ladders, and water releases from reservoirs timed to help fish spawn, or to restore more natural flow regimes

Like recreational usage, environmental usage is non-consumptive but may reduce the availability of water for other users at specific times and places. For example, water release from a reservoir to help fish spawn may not be available to farms upstream, and water retained in a river to maintain waterway health would not be available to water abstractors downstream.

3.6 WATER STRESS

The concept of water stress is relatively simple: According to the World Business Council for Sustainable Development, it applies to situations where there is not enough water for all uses, whether agricultural, industrial or domestic. Defining thresholds for stress in terms of available water per capita is more complex, however, entailing assumptions about water use and its efficiency. Nevertheless, it has been proposed that when annual per capita renewable freshwater availability is less than 1,700 cubic meters, countries begin to experience periodic or regular water stress. Below 1,000 cubic meters, water scarcity begins to hamper economic development and human health and well-being.

3.6.1 Population Growth

In 2000, the world population was 6.2 billion. The UN estimates that by 2050 there will be an additional 3.5 billion people with most of the growth in developing countries that already suffer water stress. Thus, water demand will increase unless there are corresponding increases in water conservation and recycling of this vital resource. In building on the data presented here by the UN, the World Bank goes on to explain that access to water for producing food will be one of the main challenges in the decades to come. Access to water will need to be balanced with the importance of managing water itself in a sustainable way while taking into account the impact of climate change, and other environmental and social variables.

3.6.2 Expansion of Business Activity

Business activity ranging from industrialization to services such as tourism and entertainment continues to expand rapidly. This expansion requires increased water services including both supply and sanitation, which can lead to more pressure on water resources and natural ecosystem.

3.6.3 Rapid Urbanization

The trend towards urbanization is accelerating. Small private wells and septic tanks that work well in low-density communities are not feasible within high-density urban areas. Urbanization requires significant investment in waterinfrastructure in order to deliver water to individuals and to process the concentrations of wastewater – both from individuals and from business. These polluted and contaminated waters must be treated or they pose unacceptable public health risks.

In 60% of European cities with more than 100,000 people, groundwater is being used at a faster rate than it can be replenished. Even if some water remains available, it costs more and more to capture it.

3.6.4 Climate Change

Climate change could have significant impacts on water resources around the world because of the close connections between the climate and hydrological cycle. Rising temperatures will increase evaporation and lead to increases in precipitation, though there will be regional variations in rainfall. Overall, the global supply of freshwater will increase. Bothdroughts and floods may become more frequent in different regions at different times, and dramatic changes in snowfalland snow melt are expected in mountainous areas. Higher temperatures will also affect water quality in ways that are not well understood. Possible impacts include increased eutrophication. Climate change could also mean an increase in demand for farm irrigation, garden sprinklers, and perhaps even swimming pools. There is now ample evidence that increased hydrologic variability and change in climate has and will continue have a profound impact on the water sector through the hydrologic cycle, water availability, water demand, and water allocation at the global, regional, basin, and local levels.

3.7 DEPLETION OF AQUIFERS

Due to the expanding human population, competition for water is growing such that many of the worlds major aquifers are becoming depleted. This is due both for direct human consumption as well as agricultural irrigation by groundwater. Millions of pumps of all sizes are currently extracting groundwater throughout the world. Irrigation in dry areas such as northern China, Nepal and India is supplied by groundwater, and is being extracted at an unsustainable rate. Cities that have experienced aquifer drops between 10 to 50 meters include Mexico City, Bangkok, Beijing, Madras and Shanghai.

3.7.1 Pollution and Water Protection

Water pollution is one of the main concerns of the world today. The governments of numerous countries have striven to find solutions to reduce this problem. Many pollutants threaten water supplies, but the most widespread, especially in developing countries, is the discharge of raw sewage into natural waters; this method of sewage disposal is the most common method in underdeveloped countries, but also is prevalent in quasi-developed countries such as China, India, Nepal and Iran. Sewage, sludge, garbage, and even toxic pollutants are all dumped into the water. Even if sewage is treated, problems still arise. Treated sewage forms sludge, which may be placed in landfills, spread out on land, incinerated or dumped at sea. In addition to sewage, nonpoint source pollution such as agricultural runoff is a significant source of pollution in some parts of the world, along with urban stormwater runoff and chemical wastes dumped by industries and governments.

3.8 WATER AND CONFLICTS

Competition for water has widely increased, and it has become more difficult to conciliate the necessities for water supply for human consumption, food production, ecosystems and other uses. Water administration is frequently involved in contradictory and complex problems. Approximately 10% of the worldwide annual runoff is used for human necessities. Several areas of the world are flooded, while others have such low precipitations that human life is almost impossible. As population and development increase, raising water demand, the possibility of problems inside a certain country or region increases, as it happens with others outside the region.

Over the past 25 years, politicians, academics and journalists have frequently predicted that disputes over water would be a source of future wars. Commonly cited quotes include: that of former Egyptian Foreign Minister and former Secretary-General of the United Nations Boutrous Ghali, who forecast, "The next war in the Middle East will be fought over water, not politics"; his successor at the UN, Kofi Annan, who in 2001 said, "Fierce competition for fresh water may well become a source of conflict and wars in the future," and the former Vice President of the World Bank, Ismail Serageldin, who said the wars of the next century will be over water unless significant changes in governance occurred. The water wars hypothesis had its roots in earlier research carried out on a small number of transboundary rivers such as the Indus, Jordan and Nile. These particular rivers became the focus because they had experienced water-related disputes. Specific events cited as evidence include Israel's bombing of Syria's attempts to divert the Jordan's headwaters, and military threats by Egypt against any country building dams in the upstream waters of the Nile. However, while some links made between conflict and water were valid, they did not necessarily represent the norm.

The only known example of an actual inter-state conflict over water took place between 2500 and 2350 BC between theSumerian states of Lagash and Umma. Water stress has most often led to conflicts at local and regional levels.Tensions arise most often within national borders, in the downstream areas of distressed river basins. Areas such as the lower regions of China's Yellow River or the Chao Phraya River in Thailand, for example, have already been experiencingwater stress for several years. Water stress can also exacerbate conflicts and political tensions which are not directly caused by water. Gradual reductions over time in the quality and/or quantity of fresh water can add to the instability of a region by depleting the health of a population, obstructing economic development, and exacerbating larger conflicts.

3.9 SHARED WATER RESOURCES CAN PROMOTE COLLABORATION

Water resources that span international boundaries, are more likely to be a source of collaboration and cooperation, than war. Scientists working at the International Water

Management Institute have been investigating the evidence behind water war predictions. Their findings show that, while it is true there has been conflict related to water in a handful of international basins, in the rest of the world's approximately 300 shared basins the record has been largely positive. This is exemplified by the hundreds of treaties in place guiding equitable water use between nations sharing water resources. The institutions created by these agreements can, in fact, be one of the most important factors in ensuring cooperation rather than conflict.

The International Union for the Conservation of Nature (IUCN) published the book Share: Managing water across boundaries. One chapter covers the functions of trans-boundary institutions and how they can be designed to promote cooperation, overcome initial disputes and find ways of coping with the uncertainty created by climate change. It also covers how the effectiveness of such institutions can be monitored.

3.10 WORLD WATER SUPPLY AND DISTRIBUTION

Food and water are two basic human needs. However, global coverage figures from 2002 indicate that, of every 10 people:

- Roughly 5 have a connection to a piped water supply at home (in their dwelling, plot or yard);
- 3 make use of some other sort of improved water supply, such as a protected well or public standpipe;
- 2 are unserved;
- In addition, 4 out of every 10 people live without improved sanitation.

At Earth Summit 2002 governments approved a Plan of Action to:

- Halve by 2015 the proportion of people unable to reach or afford safe drinking water. The Global Water Supply and Sanitation Assessment 2000 Report (GWSSAR) defines "Reasonable access" to water as at least 20 liters per person per day from a source within one kilometer of the user's home.
- Halve the proportion of people without access to basic sanitation. The GWSSR defines "Basic sanitation" as private or shared but not public disposal systems that separate waste from human contact.

In 2025, water shortages will be more prevalent among poorer countries where resources are limited and population growth is rapid, such as the Middle East, Africa, and parts of Asia. By 2025, large urban and peri-urban areas will require new infrastructure to provide safe water and adequate sanitation. This suggests growing conflicts with agricultural water users, who currently consume the majority of the water used by humans. Generally speaking the more developed countries of North America, Europe and Russia will not see a serious threat to water supply by the year 2025, not only because of their relative wealth, but more importantly their populations will be better aligned with available water resources. North Africa, the Middle East, South Africa and northern China will face very severe water shortages due to physical scarcity and a condition of overpopulation relative to their carrying capacity with respect to water supply. Most of South America, Sub-Saharan Africa, Southern China and India will face water supply shortages by 2025; for these latter regions the causes of scarcity will be economic constraints to developing safe drinking water, as well as excessive population growth. 1.6 billion people have gained access to a safe water source since 1990. The proportion of people in developing countries with access to safe water is calculated to have improved from 30 percent in 1970 to 71 percent in 1990, 79 percent in 2000 and 84 percent in 2004. This trend is projected to continue.

3.11 ECONOMIC CONSIDERATIONS

Water supply and sanitation require a huge amount of capital investment in infrastructure such as pipe networks, pumping stations and water treatment works. It is estimated that Organisation for Economic Co-operation and Development (OECD) nations need to invest at least USD 200 billion per year to replace aging water infrastructure to guarantee supply, reduce leakage rates and protect water quality.

International attention has focused upon the needs of the developing countries. To meet the Millennium Development Goals targets of halving the proportion of the population lacking access to safe drinking water and basic sanitation by 2015, current annual investment on the order of USD 10 to USD 15 billion would need to be roughly doubled. This does not include investments required for the maintenance of existing infrastructure.

Once infrastructure is in place, operating water supply and sanitation systems entails significant ongoing costs to cover personnel, energy, chemicals, maintenance and other expenses. The sources of money to meet these capital and operational costs are essentially either user fees, public funds or some combination of the two

3.12 DRINKING WATER QUALITY STANDARDS

Drinking water quality standards describes the quality parameters set for drinking water. Despite the truism that every human on this planet needs drinking water to survive and that water may contain many harmful constituents, there are no universally recognized and accepted international standards for drinking water. Even where standards do exist, and are applied, the permitted concentration of individual constituents may vary by as much as ten times from one set of standards to another.

Many developed countries specify standards to be applied in their own country. In Europe, this includes the European Drinking Water Directive and in the USA the United

States Environmental Protection Agency (EPA) establishes standards as required by the Safe Drinking Water Act. For countries without a legislative or administrative framework for such standards, the World Health Organisation publishes guidelines on the standards that should be achieved. China adopted its own drinking water standard GB3838-2002 (Type II) enacted by Ministry of Environmental Protection in 2002. Where drinking water quality standards do exist, most are expressed as guidelines or targets rather than requirements, and very few water standards have any legal basis or, are subject to enforcement. Two exceptions are the European Drinking Water Directive and the Safe Drinking Water Act in the USA, which require legal compliance with specific standards.

In Europe, this includes a requirement for member states to enact appropriate local legislation to mandate the directive in each country. Routine inspection and, where required, enforcement is enacted by means of penalties imposed by the European Commission on non-compliant nations. Countries with guideline values as their standards include Canada, which has guideline values for a relatively small suite of parameters, New Zealand, where there is a legislative basis, but water providers have to make "best endeavours" to comply with the standards, and Australia.

3.12.1 Range of Standards

Although drinking water standards frequently are referred to as if they are simple lists of parametric values, standards documents also specify the sampling location, sampling methods, sampling frequency, analytical methods, and laboratory accreditation AQC. In addition, a number of standards documents also require calculation to determine whether a level exceeds the standard, such as taking an average. Some standards give complex, detailed requirements for the statistical treatment of results, temporal and seasonal variations, summation of related parameters, and mathematical treatment of apparently aberrant results.

3.12.2 Australian Standards

Drinking water quality standards in Australia have been developed by the Australian Government National Health and Medical Research Council (NHMRC) in the form of the Australian Drinking Water Guidelines. These guidelines provide contaminant limits (pathogen, aesthetic, organic, inorganic, and radiological) as well as guidance on applying limits for the management of drinking water in Australian drinking water treatment and distribution.

3.12.3 European Union Standards

The following parametric standards are included in the Drinking Water Directive and are expected to be enforced by appropriate legislation in every country in the European Union. Simple parametric values are reproduced here, but in many cases the original directive also provides caveats and notes about many of the values given.

- Acrylamide 0.10 ìg/l
- Antimony 5.0 ìg/l
- Arsenic 10 ìg/l
- Benzene 1.0 ìg/l
- Benzo(a)pyrene 0.010 ìg/l
- Boron 1.0 mg/l
- Bromate 10 ìg/l
- Cadmium 5.0 ìg/l
- Chromium 50 ìg/l
- Copper 2.0 mg/l
- Cyanide 50 ìg/l
- 1,2-dichloroethane 3.0 ìg/l
- Epichlorohydrin 0.10 ìg/l
- Fluoride 1.5 mg/l
- Lead 10 ìg/l
- Mercury 1.0 ìg/l
- Nickel 20 ìg/l
- Nitrate 50 mg/l
- Nitrite 0.50 mg/l
- Pesticides 0.10 ìg/l
- Pesticides - Total 0.50 ìg/l
- Polycyclic aromatic hydrocarbons 0.10 ìg/l Sum of concentrations of specified compounds;
- Selenium 10 ìg/l
- Tetrachloroethene and Trichloroethene 10 ìg/l Sum of concentrations of specified parameters
- Trihalomethanes — Total 100 ìg/l Sum of concentrations of specified compounds
- Vinyl chloride 0.50 ìg/l

3.12.4 United States Standards

In the USA, the federal legislation controlling drinking water quality is the Safe Drinking Water Act (SDWA) which is implemented by the EPA, mainly through state or territorial primacy agencies. States and territories must implement rules that are at least as stringent as EPA's to retain primary enforcement authority (primacy) over drinking water. Many states also apply their own state-specific standards, which may be more rigorous or include additional parameters.

Standards set by the EPA in the USA are not international standards since they apply only to a single country, however, many countries look to the USA for appropriate scientific and public health guidance and may reference or adopt USA standards.

3.13 WORLD HEALTH ORGANISATION GUIDELINES

The World Health Organisation (WHO) guidelines include the following recommended limits on naturally occurring constituents that may have direct adverse health impact:

- Arsenic 10ìg/l
- Barium 10ìg/l
- Boron 2400ìg/l
- Chromium 50ìg/l
- Fluoride 1500ìg/l
- Selenium 40ìg/l
- Uranium 30ìg/l

For the potential occurrence in drinking water of artificial or synthetic pollutants generated by human activities, the following standards have been proposed:

- Cadmium 3ìg/l
- Mercury 6ìg/l For inorganic mercury

Organic species:

- Benzene 10ìg/l
- Di(2-ethylhexyl)phthalate 8 ìg/l
- 1,4-Dioxane 50ìg/l
- Edetic acid 600ìg/l
- Ethylbenzene 300 ìg/l
- Hexachlorobutadiene 0.6 ìg/l
- Nitrilotriacetic acid 200ìg/l
- Pentachlorophenol 9ìg/l
- Styrene 20ìg/l
- Tetrachloroethene 40ìg/l

3.13.1 Safe Drinking Water Act

The Safe Drinking Water Act (SDWA) is the principal federal lawin the United States intended to ensure safe drinking water for the public. Pursuant to the act, the Environmental Protection Agency(EPA) is required to set standards for drinking water quality and oversee all states, localities, and water suppliers who implement these standards.

SDWA applies to every public water system (PWS) in the United States. There are currently more than 160,000 public water systems providing water to almost all Americans at some time in their lives. The Act does not cover private wells.

The SDWA does not apply to bottled water. Bottled water is regulated by the Food and Drug Administration (FDA) under theFederal Food, Drug, and Cosmetic Act.

3.14 COMPARISON OF PARAMETRIC VALUES

(Parametric value also has a specific and different mathematical meaning)

A parametric value in this context is most commonly the concentration of a substance, e.g. 30 mg/l of Iron. It may also be a count such as 500 E. coli per litre or a statistical value such as the average concentration of copper is 2 mg/l. Many countries not only specify parametric values that may have health impacts but also specify parametric values for a range of constituents that by themselves are unlikely to have any impact on health. These include colour, turbidity, pH, and theorganoleptic (aesthetic) parameters (taste and odour).

It is possible and technically acceptable to refer to the same parameter in different ways that may appear to suggest a variation in the standard required. For example, nitrite may be measured as nitrite ion or expressed as N. A standard of "Nitrite as N" set at 1.4 mg/l equals a nitrite ion concentration of 4.6 mg/l. This is an apparent difference of nearly threefold.

The following table provides a comparison of a selection of parameters for concentrations listed by WHO, the European Union, EPA, and Ministry of Environmental Protection of China.

Indicates that no standard has been identified by editors of this article and ns indicates that no standard exists. ìg/l -> Micro grams per litre or 0.001 ppm, mg/L -> 1 ppm or 1000 ìg/l

3.15 NATIONAL PRIMARY DRINKING WATER REGULATIONS

The SDWA requires EPA to establish National Primary Drinking Water Regulations (NPDWRs) for contaminants that may cause adverse public health effects. The regulations include both mandatory levels (Maximum Contaminant Levels, or MCLs) and nonenforceable health goals (Maximum Contaminant Level Goals, or MCLGs) for each included contaminant. MCLs have additional significance because they can be used under the Superfund law as "Applicable or Relevant and Appropriate Requirements" in cleanups of contaminated sites on the National Priorities List. Federal drinking water standards are organized into six groups:

- Microorganisms
- Disinfectants
- Disinfection Byproducts
- Inorganic Chemicals
- Organic Chemicals
- Radionuclides.

3.15.1 Microorganisms

EPA has issued standards for Cryptosporidium, Giardia lamblia, Legionella, coliform bacteria and enteric viruses. EPA also requires two microorganism-related tests to indicate water quality: plate count and turbidity.

Disinfectants

EPA has issued standards for chlorine, chloramine and chlorine dioxide.

Disinfection By-Products

EPA has issued standards for bromate, chlorite, haloacetic acids and trihalomethanes.

Inorganic Chemicals

EPA has issued standards for antimony, arsenic, asbestos, barium, beryllium, cadmium, chromium, copper, cyanide,fluoride, lead, mercury, nitrate, nitrite, selenium and thallium.

"Lead Free" Plumbing Requirements

The 1986 amendments require EPA to set standards limiting the concentration of lead in public water systems, and defines "lead free" pipes as:

(1) Solders and flux containing not more than 0.2 percent lead;

(2) Pipes and pipe fittings containing not more than 8.0 percent lead; and

(3) Plumbing fittings and fixtures as defined in industry-developed voluntary standards (issued no later than August 6, 1997), or standards developed by EPA in lieu of voluntary standards.

EPA issued an initial lead and copper regulation in 1991.

Organic Chemicals

EPA has issued standards for 53 organic compounds, including benzene, dioxin (2,3,7,8-TCDD), PCBs, styrene, toluene,vinyl chloride and several pesticides.

Radionuclides

EPA has issued standards for alpha particles, beta particles and photon emitters, radium and uranium.

3.16 FUTURE STANDARDS

3.16.1 Non-Community Water Systems

Future NPDWR standards will apply to non-transient non-community water systems because of concern for the long-term exposure of a stable population. It is important to note that EPA's decision to apply future NPDWRs to non-transient non-community water systems may have a significant impact on Department of Energy facilities that operate their own drinking water systems.

3.16.2 Unregulated Contaminants

The SDWA requires EPA to identify and list unregulated contaminants which may require regulation. The Agency must periodically publish this list, called the "Contaminant Candidate List" and decide whether to regulate at least five or more listed contaminants. EPA uses this list to prioritize research and data collection efforts, which support the regulatory determination process.

3.16.3 Monitoring, Compliance and Enforcement

Public water systems are required to regularly monitor their water for contaminants. Water samples must be analyzed using EPA-approved testing methods, by laboratories that are certified by EPA or a state agency.

Community water systems—those systems that serve the same people throughout the year—must provide an annual "Consumer Confidence Report" to customers. The report identifies contami-nants, if any, in the drinking water and explains the potential health impacts. Oversight of public water systems is managed by "primacy" agencies, which are either state government agencies or EPA regional offices. All state and territories, except Wyoming and the District of Columbia, have received primacy approval from EPA, to supervise the PWS in their respective jurisdictions. A PWS is required to submit periodic monitoring reports to its primacy agency. Violations of SDWA requirements are enforced initially through a primacy agency's notification to the PWS, and if necessary following up with formal orders and fines.

3.17 RELATED PROGRAMS

3.17.1 Airline Water Supplies

In 2004, EPA tested drinking water quality on commercial aircraft and found that 15 percent of tested aircraft water systems tested positive for total coliform bacteria. EPA published a final regulation for aircraft public water systems in 2009. The regulation requires air carriers operating in the U.S. to conduct coliform sampling, management practices, corrective action, public notification, operator training, and reporting and recordkeeping. An airline with a non-complying aircraft must restrict public access to the on-board water system for a specified period.

3.17.2 Underground Injection Control (UIC) Program

The 1974 act authorized EPA to regulate injection wells in order to protect underground sources of drinking water.Congress amended the SDWA in 2005 to exclude hydraulic fracturing, an industrial process for recovering oil and natural gas, from coverage under the UIC program. This exclusion has been called the "Halliburton Loophole". Halliburton is the world's largest provider of hydraulic fracturing services.

The Fracturing Responsibility and Awareness of Chemicals Act (H.R. 2766, S. 1215) would have repealed the exemption for hydraulic fracturing in the SDWA and regulated the oil and natural gas recovery process under the UIC program. The bill was introduced in the 111th Congress (2009) and did not pass. A similar amendment was introduced in January 2015 as a rider on the Keystone XL Pipeline authorization, but it was rejected.

3.17.3 Whistleblower Protection

The SDWA includes a whistleblower protection provision. Employees in the US who believe they were fired or suffered another adverse action related to enforcement of this law have 30 days to file a written complaint with the Occupational Safety and Health Administration.

3.18 HISTORY OF WATER TREATMENT

3.18.1 Prelude

Prior to the SDWA there were few national enforceable requirements for drinking water. Improvements in testing were allowing the detection of smaller concentrations of contaminant and allowing more tests to be run. Many states had drinking water regulations prior to adoption of the federal SDWA.

1974 Act: The Safe Drinking Water Act was one of several pieces of environmental legislation in the 1970s. Discovery of organic contamination in public drinking water and the lack of enforceable, national standards persuaded Congress to take action.

1986 Amendments: The 1986 SDWA amendments required EPA to apply future NPDWRs to both community and non-transient non-community water systems when it evaluated and revised current regulations. The first case in which this was applied was the "Phase I" final rule, published on July 8, 1987. At that time NPDWRs were promulgated for certain synthetic volatile organic compounds and applied to non-transient non-community water systems as well as community water systems. This rulemaking also clarified that non-transient non-community water systems were not subject to MCLs that were promulgated before July 8, 1987. The 1986 amendments were signed into law by President Ronald Reagan on June 19, 1986.

In addition to requiring more contaminants to be regulated, the 1986 amendments included:

- Well head protection
- New monitoring for certain substances
- Filtration for certain surface water systems
- Disinfection for certain groundwater systems
- Restriction on lead in solder and plumbing
- More enforcement powers.

3.19 1996 SDWA AMENDMENTS

In 1996, Congress amended the Safe Drinking Water Act to emphasize sound science and risk-based standard setting, small water supply system flexibility and technical assistance, community-empowered source water assessment and protection, public right-to-know, and water system infrastructure assistance through a multi-billion-dollar state revolving loan fund. The amendments were signed into law by President Bill Clinton on August 6, 1996.

3.19.1 Main Points of the 1996 Amendments

1. **Consumer Confidence Reports:** All community water systems must prepare and distribute annual reports about the water they provide, including information on detected contaminants, possible health effects, and the water's source.
2. **Cost-Benefit Analysis:** EPA must conduct a thorough cost-benefit analysis for every new standard to determine whether the benefits of a drinking water standard justify the costs.
3. **Drinking Water State Revolving Fund:** States can use this fund to help water systems make infrastructure or management improvements or to help systems assess and protect their source water.
4. **Microbial Contaminants and Disinfection Byproducts:** EPA is required to strengthen protection for microbial contaminants, including cryptosporidium, while strengthening control over the byproducts of chemical disinfection. EPA promulgated the Stage 1 Disinfectants and Disinfection Byproducts Rule and the Interim Enhanced Surface Water Treatment Rule to address these risks.
5. **Operator Certification:** Water system operators must be certified to ensure that systems are operated safely. EPA issued guidelines in 1999 specifying minimum standards for the certification and recertification of the operators of community

and non-transient, noncommunity water systems. These guidelines apply to state operator certification programs. All states are currently implementing EPA-approved operator certification programs.

6. **Public Information and Consultation:** SDWA emphasizes that consumers have a right to know what is in their drinking water, where it comes from, how it is treated, and how to help protect it. EPA distributes public information materials (through its Drinking Water Hotline, Safewater web site, and Resource Center) and holds public meetings, working with states, tribes, water systems, and environmental and civic groups, to encourage public involvement.
7. **Small Water Systems:** Small water systems are given special consideration and resources under SDWA, to make sure they have the managerial, financial, and technical ability to comply with drinking water standards.

3.20 2005 AMENDMENT

Through the Energy Policy Act of 2005, the Safe Drinking Water Act was amended to exclude the underground injection of any fluids or propping agents other than diesel fuels used in hydraulic fracturing operations from being considered as "underground injections" for the purposes of the law.

3.21 PROPOSED AMENDMENTS

Controversy surrounds the practice of hydraulic fracturing as a threat to drinking water supplies. The Fracturing Responsibility and Awareness of Chemicals Act, dubbed the "FRAC Act," was introduced to both houses of the 111th Congress on June 9, 2009. The bill would have repealed the exemption for hydraulic fracturing in the SDWA and regulated the oil and natural gas recovery process under the UIC program. It would have required the energy industry to disclose the chemicals it mixes with the water and sand it pumps underground in the process (also known as "fracking"), information that has largely been protected as proprietary or trade secrets. Without knowing the identity of the proprietary components, regulators cannot test for their presence. This prevents government regulators from establishing baseline levels of the substances prior to hydraulic fracturing and documenting changes in these levels, thereby making it impossible to determine whether hydraulic fracturing is contaminating the environment with these substances. The natural gas industry opposed the legislation. The House bill was introduced by representatives Diana DeGette (D-CO), Maurice Hinchey(D-NY), and Jared Polis (D-CO). The Senate version was introduced by senators Bob Casey (D-PA), and Chuck Schumer(D-NY). After a delay, the FRAC Act was re-introduced in both houses of the 112th United States Congress. In the Senate, Sen. Bob Casey (D-PA) introduced S. 587 on March 15, 2011. In the House, Rep. Diana DeGette (D-CO) introduced H.R. 1084 on March 24, 2011.

As of May 2012, Congress had not yet passed either of the FRAC Act bill

4

STRUCTURE OF AQUATIC SYSTEM

4.1 OBJECTIVES

After studying this chapter you will be able to understand:

- Aquatic ecosystem.
- Abiotic characteristics.
- Autotrophic organisms.
- Ecosystem ecology.
- Nitrogen compounds.
- Exposure systems.

4.2 INTRODUCTION

An aquatic ecosystem is an ecosystem in and surrounding a body of water, in contrast to land-based terrestrial ecosystems. Aquatic ecosystems contain communities of organisms that are dependent on each other and on their environment. The two main types of aquatic ecosystems are marine ecosystems and freshwater ecosystems. Freshwater ecosystems may be lentic (slow moving water, including pools, ponds, and lakes); lotic (faster moving water, for example streams and rivers); and wetlands (areas where the soil is saturated or inundated for at least part of the time).

The health of an aquatic ecosystem is degraded when the ecosystem's ability to absorb a stress has been exceeded. A stress on an aquatic ecosystem can be a result of physical, chemical or biological alterations to the environment. Physical alterations include changes in water temperature, water flow and light availability. Chemical alterations include changes in the loading rates of biostimulatory nutrients, oxygen-consuming materials, and toxins. Biological alterations include over-harvesting of commercial species and the introduction of exotic species. Human populations can impose excessive stresses on aquatic ecosystems.

There are many examples of excessive stresses with negative consequences. The environmental history of the Great Lakes of North America illustrates this problem, particularly how multiple stresses, such as water pollution, over-harvesting and invasive species can combine. The Norfolk Broadlands in England illustrate similar decline with pollution and invasive species. Lake Pontchartrain along the Gulf of Mexico illustrates the negative effects of different stresses including levee construction, logging of swamps, invasive species and salt water intrusion.

An ecosystem is defined as a functional unit wherein all living organisms interact with their surroundings and one another to sustain themselves in the environment. In a broad sense, an ecosystem can be categorized as a land/terrestrial ecosystem or a water/aquatic ecosystem.

Many lives are undoubtedly supported by water. Furthermore, aquatic creatures are those that can thrive in water. They also rely on water for food, shelter, reproduction, and a variety of other life functions. This particular article will offer you a fair idea of the types of aquatic ecosystems and their importance in brief!

4.3 AQUATIC ECOSYSTEM

The aquatic ecosystem definition states it is a water-based environment, wherein, living organisms interact with both physical and chemical features of the environment. These living creatures whose food, shelter, reproduction, and other essential activities depend on a water-based environment are known as aquatic organisms.

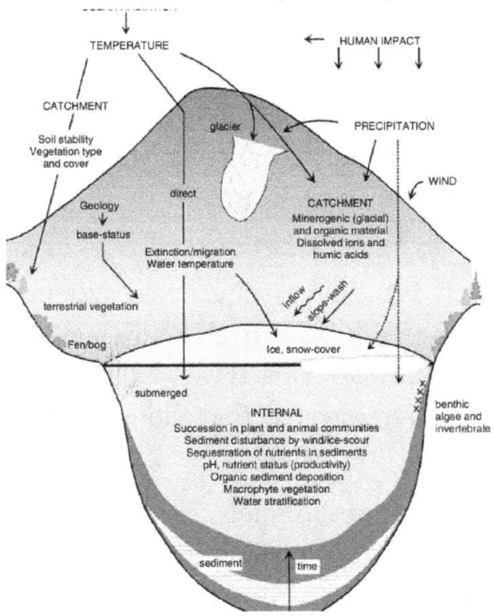

Fig. 1. Freshwater environments include lakes, ponds, rivers, seas and streams, wetlands, swamps, and so on, while marine habitats include oceans, intertidal zones, reefs, and the seabed. Water-dependent living creatures, such as animals, plants, and bacteria, call the aquatic ecosystem home.

Water plays a significant role in the management of world-scale ecosystem processes in aquatic systems, connecting the atmosphere, lithosphere, and biosphere by transferring material between them and allowing chemical reactions to occur. Water has unique physicochemical features that reflect the water body's quality. The physicochemical characteristics of an aquatic ecosystem determine how well it functions and how long it can support life forms. In the same way as sediments in terrestrial ecosystems provide substrate, nutrients, and a home for live aquatic resources, sediments in aquatic ecosystems are equivalent to the soil in terrestrial ecosystems. Sediments are significant catalysts in environmental food cycles and the two water quality dynamics.

The quality of sediment has a direct or indirect impact on the functioning of an aquatic ecosystem. The many physicochemical properties of sediment determine its quality. Similarly, the biotic mix of an aquatic environment determines how well it functions. In the aquatic environment, they serve as a trophic level and a source of energy. Fish have a significant ecological role in the whole food web at the trophic level.

Some of the most common aquatic organisms are – nekton, plankton, and benthos. Additionally, lakes, oceans, ponds, rivers, swamps, coral reefs, wetlands, etc. are a few popular aquatic ecosystem examples.

4.3.1 Marine

Aquatic Ecosystem is defined as a water-based habitat in which living species interact with the environment's physical and chemical properties. Aquatic organisms are living species that rely on a water-based environment for their food, habitat, reproduction, and other critical functions.

Nekton, plankton, and benthos are some of the most prevalent aquatic creatures. Lakes, oceans, ponds, rivers, swamps, coral reefs, wetlands, and other popular aquatic ecosystem examples include lakes, oceans, ponds, rivers, swamps, coral reefs, wetlands, and so on.

This particular ecosystem is the largest aquatic ecosystem and covers over 70% of the earth's total surface. This ecosystem is relatively more concentrated in terms of salinity. Nonetheless, the body of aquatic organisms is well-adjusted to saline water, and they may find it challenging to survive in freshwater. The following categories comprise the marine ecosystem.

Marine ecosystems cover approximately 71% of the Earth's surface and contain approximately 97% of the planet's water. They generate 32% of the world's net primary production. They are distinguished from freshwater ecosystems by the presence of dissolved compounds, especially salts, in the water. Approximately 85% of the dissolved materials in seawaterare sodium and chlorine. Seawater has an average salinity of 35 parts per thousand (ppt) of water. Actual salinity varies among different marine ecosystems.

Marine ecosystems can be divided into many zones depending upon water depth and shoreline features. The oceanic zone is the vast open part of the ocean where animals such as whales, sharks, and tuna live. The benthic zone consists of substrates below water where many invertebrates live. The intertidal zone is the area between high and low tides; in this figure it is termed the littoral zone. Other near-shore (neritic) zones can include estuaries, salt marshes, coral reefs, lagoonsand mangrove swamps. In the deep water, hydrothermal vents may occur where chemosynthetic sulfur bacteria form the base of the food web. Classes of organisms found in marine ecosystems include brown algae, dinoflagellates, corals, cephalopods, echinoderms, and sharks. Fishes caught in marine ecosystems are the biggest source of commercial foods obtained from wild populations.

Environmental problems concerning marine ecosystems include unsustainable exploitation of marine resources (for example overfishing of certain species), marine pollution, climate change, and building on coastal areas.

4.3.2 Freshwater

Freshwater ecosystems cover 0.80% of the Earth's surface and inhabit 0.009% of its total water. They generate nearly 3% of its net primary production. Freshwater ecosystems contain 41% of the world's known fish species.

There are three basic types of freshwater ecosystems:

- Lentic: slow moving water, including pools, ponds, and lakes.
- Lotic: faster moving water, for example streams and rivers.
- Wetlands: areas where the soil is saturated or inundated for at least part of the time.

4.3.3 Lentic

It includes standing water bodies like ponds and lakes and is a home to both floating and rooted plants, algae, and invertebrates. All standing water habitats, such as lakes and ponds, are included in lentic ecosystems. Algae, rooted and floating-leaved plants, and crustaceans like crabs and shrimp live in these habitats. Frogs and salamanders, as well as reptiles like alligators and water snakes, can be found here. Salamanders, frogs, water snakes, and alligators are commonly found in lentic ecosystems.

Lake ecosystems can be divided into zones. One common system divides lakes into three zones. The first, the littoral zone, is the shallow zone near the shore. This is where rooted wetland plants occur. The offshore is divided into two further zones, an open water zone and a deep water zone. In the open water zone (or photic zone) sunlight supports photosynthetic algae, and the species that feed upon them. In the deep water zone, sunlight is not available and the food web is based on detritus entering from the littoral and photic zones. Some systems use other names. The off shore areas may be called the pelagic zone, and the aphotic zone may be called the profundal zone. Inland from the littoral zone one can also frequently identify a riparian zone which has plants still affected by the presence of the lake—this can include effects from windfalls, spring flooding, and winter ice damage. The production of the lake as a whole is the result of production from plants growing in the littoral zone, combined with production from plankton growing in the open water. Wetlands can be part of the lentic system, as they form naturally along most lakeshores, the width of the wetland and littoral zone being dependent upon the slope of the shoreline and the amount of natural change in water levels, within and among years. Often dead trees accumulate in this zone, either from windfalls on the shore or logs transported to the site during floods. This woody debris provides important habitat for fish and nesting birds, as well as protecting shorelines from erosion.

Two important subclasses of lakes are ponds, which typically are small lakes that intergrade with wetlands, and waterreservoirs. Over long periods of time, lakes, or bays within them, may gradually become enriched by nutrients and slowly fill in with organic sediments, a process called succession. When humans use the watershed, the volumes of sediment entering the lake can accelerate this process. The addition of sediments and nutrients to a lake is known as eutrophication.

4.3.4 Ponds

Ponds are small bodies of freshwater with shallow and still water, marsh, and aquatic plants. They can be further divided into four zones: vegetation zone, open water, bottom mud and surface film. The size and depth of ponds often varies greatly with the time of year; many ponds are produced by spring flooding from rivers. Food webs are based both on free-floating algae and upon aquatic plants. There is usually a diverse array of aquatic life, with a few examples including algae, snails, fish, beetles, water bugs, frogs, turtles, otters and muskrats. Top predators may include large fish, herons, or alligators. Since fish are a major predator upon amphibian larvae, ponds that dry up each year, thereby killing resident fish, provide important refugia for amphibian breeding. Ponds that dry up completely each year are often known as vernal pools. Some ponds are produced by animal activity, including alligator holes and beaver ponds, and these add important diversity to landscapes.

4.3.5 Lotic

The major zones in river ecosystems are determined by the river bed's gradient or by the velocity of the current. Faster moving turbulent water typically contains greater concentrations of dissolved oxygen, which supports greater biodiversity than the slow moving water of pools. These distinctions form the basis for the division of rivers into upland and lowlandrivers. The food base of streams within riparian forests is mostly derived from the trees, but wider streams and those that lack a canopy derive the majority of their food base from algae. Anadromous fish are also an important source of nutrients. Environmental threats to rivers include loss of water, dams, chemical pollution and introduced species. A dam produces negative effects that continue down the watershed. The most important negative effects are the reduction of spring flooding, which damages wetlands, and the retention of sediment, which leads to loss of deltaic wetlands.

4.3.6 Wetlands

Wetlands are dominated by vascular plants that have adapted to saturated soil. There are four main types of wetlands: swamp, marsh, fen and bog (both fens and bogs are types of mire). Wetlands are the most productive natural ecosystems in the world because of the proximity of water and soil. Hence they support large numbers of plant and animal species. Due to their productivity, wetlands are often converted into dry land with dykes and drains and used for agricultural purposes. The construction of dykes, and dams, has negative consequences for individual wetlands and entire watersheds. Their closeness to lakes and rivers means that they are often developed for human settlement. Once settlements are constructed and protected by dykes, the settlements then become vulnerable to land subsidence and ever increasing risk of flooding. The Louisiana coast around New Orleans is a well-known example; the Danube Delta in Europe is another.

4.3.7 Functions

Aquatic ecosystems perform many important environmental functions. For example, they recycle nutrients, purify water, attenuate floods, recharge ground water and provide habitats for wildlife. Aquatic ecosystems are also used for human recreation, and are very important to the tourism industry, especially in coastal regions.

The health of an aquatic ecosystem is degraded when the ecosystem's ability to absorb a stress has been exceeded. A stress on an aquatic ecosystem can be a result of physical, chemical or biological alterations of the environment. Physical alterations include changes in water temperature, water flow and light availability. Chemical alterations include changes in the loading rates of biostimulatory nutrients, oxygen consuming materials, and toxins. Biological alterations include over-harvesting of commercial species and the introduction of exotic species. Human populations can impose excessive stresses on aquatic ecosystems. There are many examples of excessive stresses with negative consequences. Consider three. The environmental history of the Great Lakes of North America illustrates this problem, particularly how multiple stresses, such as water pollution, over-harvesting and invasive species can combine. The Norfolk Broadlands in England illustrate similar decline with pollution and invasive species. Lake Pontchartrain along the Gulf of Mexico illustrates the negative effects of different stresses including levee construction, logging of swamps, invasive species and salt water intrusion.

4.4 ABIOTIC CHARACTERISTICS

An ecosystem is composed of biotic communities that are structured by biological interactions and abiotic environmental factors. Some of the important abiotic environmental factors of aquatic ecosystems include substrate type, water depth, nutrient levels, temperature, salinity, and flow. It is often difficult to determine the relative importance of these factors without rather large experiments. There may be complicated feed back loops. For example, sediment may determine the presence of aquatic plants, but aquatic plants may also trap sediment, and add to the sediment through peat.

The amount of dissolved oxygen in a water body is frequently the key substance in determining the extent and kinds of organic life in the water body. Fish need dissolved oxygen to survive, although their tolerance to low oxygen varies among species; in extreme cases of low oxygen some fish even resort to air gulping. Plants often have to produce aerenchyma, while the shape and size of leaves may also be altered. Conversely, oxygen is fatal to many kinds of anaerobicbacteria.

Nutrient levels are important in controlling the abundance of many species of algae. The relative abundance of nitrogen and phosphorus can in effect determine which species of algae come to dominate. Algae are a very important source of food for aquatic life, but at the same time, if they become over-abundant, they can cause declines in fish when they decay. Similar over-abundance of algae in coastal environments such as the Gulf of Mexico produces, upon decay, a hypoxic region of water known as a dead zone.

The salinity of the water body is also a determining factor in the kinds of species found in the water body. Organisms in marine ecosystems tolerate salinity, while many freshwater organisms are intolerant of salt. The degree of salinity in an estuary or delta is an important control upon the type of wetland (fresh, intermediate, or brackish), and the associated animal species. Dams built upstream may reduce spring flooding, and reduce sediment accretion, and may therefore lead to saltwater intrusion in coastal wetlands.

Freshwater used for irrigation purposes often absorbs levels of salt that are harmful to freshwater organisms.

4.5 BIOTIC CHARACTERISTICS

The biotic characteristics are mainly determined by the organisms that occur. For example, wetland plants may produce dense canopies that cover large areas of sediment—or snails or geese may graze the vegetation leaving large mud flats. Aquatic environments have relatively low oxygen levels, forcing adaptation by the organisms found there. For example, many wetland plants must produce aerenchyma to carry oxygen to roots. Other biotic characteristics are more subtle and difficult to measure, such as the relative importance of competition, mutualism or predation. There are a growing number of cases where predation by coastal herbivores including snails, geese and mammals appears to be a dominant biotic factor.

4.6 AUTOTROPHIC ORGANISMS

Autotrophic organisms are producers that generate organic compounds from inorganic material. Algae use solar energy to generate biomass from carbon dioxide and are possibly the most important autotrophic organisms in aquatic environments. Of course, the more shallow the water, the greater the biomass contribution from rooted and floating vascular plants. These two sources combine to produce the extraordinary production of estuaries and wetlands, as this autotrophic biomass is converted into fish, birds, amphibians and other aquatic species.

Chemosynthetic bacteria are found in benthic marine ecosystems. These organisms are able to feed on hydrogen sulfide in water that comes from volcanic vents. Great concentrations of animals that feed on these bacteria are found around volcanic vents. For example, there are giant tube worms (Riftia pachyptila) 1.5 m in length and clams (Calyptogena magnifica) 30 cm long.

4.7 HETEROTROPHIC ORGANISMS

Heterotrophic organisms consume autotrophic organisms and use the organic compounds in their bodies as energy sources and as raw materials to create their own biomass. Euryhaline organisms are salt tolerant and can survive in marine ecosystems, while stenohaline or salt intolerant species can only live in freshwater environments

4.8 ECOSYSTEM ECOLOGY

Ecosystem ecology is the integrated study of living (biotic) and non-living (abiotic) components of ecosystems and their interactions within an ecosystem framework. This science examines how ecosystems work and relates this to their components such as chemicals, bedrock, soil, plants, and animals.

Ecosystem ecology examines physical and biological structures and examines how these ecosystem characteristics interact with each other. Ultimately, this helps us understand how to maintain high quality water and economically viable commodity production. A major focus of ecosystem ecology is on functional processes, ecological mechanisms that maintain the structure and services produced by ecosystems. These includeprimary productivity (production of biomass), decomposition, and trophicinteractions.

Studies of ecosystem function have greatly improved human understanding of sustainable production of forage, fiber, fuel, and provision of water. Functional processes are mediated by regional-to-local level climate, disturbance, and management. Thus ecosystem ecology provides a powerful framework for identifying ecological mechanisms that interact with global environmental problems, especially global warming and degradation of surface water.

This example demonstrates several important aspects of ecosystems:

1. Ecosystem boundaries are often nebulous and may fluctuate in time
2. Organisms within ecosystems are dependent on ecosystem level biological and physical processes

3. Adjacent ecosystems closely interact and often are interdependent for maintenance of community structure and functional processes that maintain productivity and biodiversity

These characteristics also introduce practical problems into natural resource management. Who will manage which ecosystem? Will timber cutting in the forest degrade recreational fishing in the stream? These questions are difficult for land managers to address while the boundary between ecosystems remains unclear; even though decisions in one ecosystem will affect the other. We need better understanding of the interactions and interdependencies of these ecosystems and the processes that maintain them before we can begin to address these questions.

Ecosystem ecology is an inherently interdisciplinary field of study. An individual ecosystem is composed of populations oforganisms, interacting within communities, and contributing to the cycling of nutrients and the flow of energy. The ecosystem is the principal unit of study in ecosystem ecology.

Population, community, and physiological ecology provide many of the underlying biological mechanisms influencing ecosystems and the processes they maintain. Flowing of energy and cycling of matter at the ecosystem level are often examined in ecosystem ecology, but, as a whole, this science is defined more by subject matter than by scale. Ecosystem ecology approaches organisms and abiotic pools of energy and nutrients as an integrated system which distinguishes it from associated sciences such as biogeochemistry. Biogeochemistry and hydrology focus on several fundamental ecosystem processes such as biologically mediated chemical cycling of nutrients and physical-biological cycling of water. Ecosystem ecology forms the mechanistic basis for regional or global processes encompassed by landscape-to-regional hydrology, global biogeochemistry, and earth system science.

4.8.1 History of Ecosystem Ecology

Ecosystem ecology is philosophically and historically rooted in terrestrial ecology. The ecosystem concept has evolved rapidly during the last 100 years with important ideas developed by Frederic Clements, a botanist who argued for specific definitions of ecosystems and that physiological processes were responsible for their development and persistence.Although most of Clements ecosystem definitions have been greatly revised, initially by Henry Gleason and Arthur Tansley, and later by contemporary ecologists, the idea that physiological processes are fundamental to ecosystem structure and function remains central to ecology.

Later work by Eugene Odum and Howard T. Odum quantified flows of energy and matter at the ecosystem level, thus documenting the general ideas proposed by Clements and his contemporary Charles Elton.

In this model, energy flows through the whole system were dependent on biotic and abiotic interactions of each individual component (species, inorganic pools of nutrients, etc.). Later work demonstrated that these interactions and flows applied to nutrient cycles, changed over the course of succession, and held powerful controls over ecosystem productivity. Transfers of energy and nutrients are innate to ecological systems regardless of whether they are aquatic or terrestrial. Thus, ecosystem ecology has emerged from important biological studies of plants, animals, terrestrial, aquatic, and marine ecosystems.

4.8.2 Ecosystem Services

Ecosystem services are ecologically mediated functional processes essential to sustaining healthy human societies. Water provision and filtration, production of biomass in forestry, agriculture, and fisheries, and removal of greenhouse gases such as carbon dioxide (CO_2) from the atmosphere are examples of ecosystem services essential to public healthand economic opportunity. Nutrient cycling is a process fundamental to agricul-tural and forest production.

However, like most ecosystem processes, nutrient cycling is not an ecosystem characteristic which can be "dialed" to the most desirable level. Maximizing production in degraded systems is an overly simplistic solution to the complex problems of hunger and economic security. For instance, intensive fertilizer use in the midwestern United States has resulted in degraded fisheries in the Gulf of Mexico. Regrettably, a "Green Revolution" of intensive chemical fertilization has been recommended for agriculture in developed and developing countries. These strategies risk alteration of ecosystem processes that may be difficult to restore, especially when applied at broad scales without adequate assessment of impacts. Ecosystem processes may take many years to recover from significant disturbance.

For instance, large-scale forest clearance in the northeastern United States during the 18th and 19th centuries has altered soil texture, dominant vegetation, and nutrient cycling in ways that impact forest productivity in the present day. An appreciation of the importance of ecosystem function in maintenance of productivity, whether in agriculture or forestry, is needed in conjunction with plans for restoration of essential processes. Improved knowledge of ecosystem function will help to achieve long-term sustainability and stability in the poorest parts of the world.

4.9 APPLICATIONS AND IMPORTANCE

4.9.1 Lessons from Two Central American Cities

The biosphere has been greatly altered by the demands of human societies. Ecosystem ecology plays an important role in understanding and adapting to the most pressing current environmental problems. Restoration ecology and ecosystem management are closely associated with ecosystem ecology. Restoring highly degraded resources depends on integration of functional mechanisms of ecosystems.

Without these functions intact, economic value of ecosystems is greatly reduced and potentially dangerous conditions may develop in the field. For example, areas within the mountainous western highlands of Guatemala are more susceptible to catastrophic landslides and crippling seasonal water shortages due to loss of forest resources. In contrast, cities such asTotonicapán that have preserved forests through strong social institutions have greater local economic stability and overall greater human well-being.

This situation is striking considering that these areas are close to each other, the majority of inhabitants are of Mayandescent, and the topography and overall resources are similar. This is a case of two groups of people managing resources in fundamentally different ways. Ecosystem ecology provides the basic science needed to avoid degradation and to restore ecosystem processes that provide for basic human need.

4.10 FRESHWATER ECOSYSTEM

Freshwater ecosystems are a subset of Earth's aquatic ecosystems. They include lakes and ponds, rivers, streams, springs, and wetlands. They can be contrasted with marine ecosystems, which have a larger salt content. Freshwater habitats can be classified by different factors, including temperature, light penetration, and vegetation.

Freshwater ecosystems can be divided into lentic ecosystems (still water) andlotic ecosystems (flowing water).

Limnology (and its branch freshwater biology) is a study about freshwater ecosystems. It is a part of hydrobiology.

Original efforts to understand and monitor freshwater ecosystems were spurred on by threats to human health (ex. Cholera outbreaks due to sewage contamination). Early monitoring focussed on chemical indicators, then bacteria, and finally algae, fungi and protozoa. A new type of monitoring involves differing groups of organisms (macroinvertebrates, macrophytes and fish) and the stream conditions associated with them.

Current biomonitering techniques focus mainly on community structure or biochemical oxygen demand. Responses are measured by behavioural changes, altered rates of growth, reproduction or mortality. Macroinvertebrates are most often used in these models because of well known taxonomy, ease of collection, sensitivity to a range of stressors, and their overall value to the ecosystem. Most of these measurements are difficult to extrapolate on a large scale, however.

The use of reference sites is common when assessing what a healthy freshwater ecosystem should "look like". Reference sites are easier to reconstruct in standing water than moving water. Preserved indicators such as diatom valves, macrophyte pollen, insect chitin and fish scales can be used to establish a reference ecosystem representative of a time before large scale human disturbance.

Common chemical stresses on freshwater ecosystem health include acidification, eutrophication and copper and pesticide contamination.

4.11 EXTINCTION OF FRESHWATER FAUNA

Over 123 freshwater fauna species have gone extinct in North America since 1900. Of North American freshwater species, an estimated 48.5% of mussels, 22.8% of gastropods, 32.7% of crayfishes, 25.9% of amphibians, and 21.3% of fishes are either endangered or threatened. Extinction rates of many species may increase severely into the next century because of invasive species, loss of keystone species, and species which are already functionally extinct. Projected extinction rates for freshwater animals are around five times greater than for land animals, and are comparable to the rates for rainforest communities. Recent extinction trends can be attributed largely to sedimentation, stream fragmentation, chemical and organic pollutants, dams, and invasive species.

4.11.1 Physical and Chemical Factors

Morphometry refers to physical factors (shape, size, structure, etc) that determine the lake basin. Generally, lakes that are small in surface area and larger in depth exhibit higher water quality than those that are larger in surface area and shallow in depth. This is referred to as surface to volume ratio (S/V).

Maximum depth and median depth along with percentage of littoral area (part of the lake basin where enough light penetrates to allow growth of plants and algae) is important in the lake's morphometry. If a lake has a very large watershed area compared to its surface area, it will receive larger surface runoff with substantial nutrient loads. Depth analyses, including sediment measurement at various depths, volumes of strata and shoreline characteristics are often critical to the investigation of biological, chemical and physical properties of fresh waters. Morphometric parameters are needed to evaluate erosion, nutrient loading rates, chemical mass, heat content, thermal stability, biological productivity and effectiveness of growth, and many other structural and functional components of the ecosystem. Management techniques such as loading capacity for effluents and selective removal of undesirable components of the biota are also dependent on detailed knowledge of the morphometry and flow characteristics.

The water in lakes is balanced by the basic hydrological relationship in which change in water storage is governed by inputs from all sources minus water losses. Water income from precipitation, surface influents and groundwater sources is balanced by outflows from surface effluents, seepage to groundwater and evapotranspiration. Each of these inflows and outflows vary seasonally and geographically and is governed by the characteristics of particular lake basins, their groundwater, drainage basins and climate. The hydrological cycle, which can be altered extensively by human induced changes of surface water systems, determines the distribution of lakes in relation to the suitable catchment (watershed) area.

4.12 WATERSHED

The lowest area of the catchment where water collects to form a lake is called lake basin. The region from where water drains into the lake basin is known as catchment area or drainage basin of the lake. Water may directly drain into the lake or pass through a system of streams and rivers, and other depressions on the way to the lake. The spatial extent of the catchment and the amount of rain falling on it ultimately determine the volume of water entering the lakes.

Watershed contributes surface-runoff to the water body. Watershed on a defined drainage channel includes all the land and water areas that drain. It is marked by an elevated line that forms a division between two areas drained by separate stream or bodies of water, such as a small watershed of a few hectares that drains into a small stream forming part of a larger watershed that in turn forms a part of a still larger watershed until the combined watershed becomes a major river basin draining millions of square kilometres of land. There can be sub watersheds within a watershed like a tributary to a lake having its own watershed, which is a part of the larger total drainage area to the lake. A lake reflects the watershed size, topography, geology, land use, soil fertility, erodibility and vegetation.

Water quality decreases with increasing ratio of watershed area to lake area. In a large watershed, water from precipitation will interact and leach minerals from the soil before being discharged into the lake. Lakes with small watersheds, maintained primarily by groundwater flow, are known as seepage lakes. Where as, lakes fed by inflowing streams or rivers are known as drainage lakes.

Land use has an impact on the quantity and quality of water entering into the lake. In urban areas, the high proportion of impervious surfaces prevents the seepage of rainwater into the soil increasing the rate of surface water flow to the lake. High flushing rates from urban areas can increase erosion of the watershed area providing sufficient force to carry materials (minerals, etc) along with the soil affecting the water quality.

Soil types and their susceptibility to erosion is another important component of the watershed. The type and abundance of vegetation determines the extent of erosion. Areas with native undisturbed vegetation are less prone to erosion than areas with disturbed vegetation like agricultural lands.

Another aspect of the watershed to be considered is whether the lake is a closed (no outlet) or open (with outlet) system. In the closed system, all the water draining in stays, not flowing into a channel or river, making it more sensitive than an open system. The upstream water quality has an important role in the health of the lake ecosystem. Lakes in the upper end of the watershed may have a different water quality than the downstream. Depending on the physical and biological factors of the upstream lakes, the downstream water quality may be affected either positively or negatively.

Factors Influencing Watershed Operations:

(a) ***Size:*** Both run-off volume and rate increases with increase in watershed size. However, both rate and volume per unit of watershed area decreases as the area increases. The size or area of watershed is an important parameter in determining the peak rate of run-off.

(b) ***Shape:*** Long and narrow watersheds have longer times of concentration resulting in lower runoff-rates than more square watersheds of similar size, which have a number of tributaries discharging into the main channel. This type of concentration also affects the amount of water infiltrating into the soil in the watershed. The longer time it takes to leave the watershed, the greater is its seepage into the soil.

(c) ***Land-slope:*** Slope has major implications on land-use. The speed and extent of run-off depend on the slope of the land. Greater the slope, greater is the velocity of flow of run-off water. If velocity is doubled, energy and consequently erosion also increases. The degree of slope sets limits on land use for annual crops, plantation and land reclamation, depending on soil depth.

(d) ***Drainage pattern:*** Drainage pattern of an area depends on the course of the streams and their tributaries. Land-slope, lithology and structure influence the drainage pattern. In general, coarser the drainage texture, higher is the conductivity. Finer drainage texture results in heavier soil type. Drainage patterns act as guidelines to locate vulnerable areas requiring different kinds and degrees of soil conservation measures.

(e) ***Soil and geology:*** Soil and geology of the watershed also determine the amount of water percolating into the ground. Soil character also determines the amount of silt that will be washed down into water harvesting structures.

(f) ***Vegetative cover:*** The type and quality of vegetative cover of the watershed influence run-off, infiltration rates, erosion, sediment production and evapotranspiration rate. Dense vegetation reduces erosion.

(g) ***Precipitation:*** Amount and nature of precipitation is an important factor determining the rate of run-off into the water body. Rainfall distributed evenly throughout the year has a different impact than a sudden sharp seasonal rainfall.

4.13 WATER BUDGET IN LAKES

The water budget of lakes provides useful information about the availability of water in lakes at any time. The basic parameters of the water budget may be classified as recharge and discharge. These can be further divided as inflow from precipitation, surface runoff and groundwater and outflow from river and evaporation. Hydrogeological, hydromorphological and hydrodynamic conditions influence the water budget in the lake basin.

The water budget equation of a lake can be expressed as follows:

$$D\,S = I_s + I_u + P_l - Q_s - Q_u - E_l$$

where D S = change in water storage, Is = Surface inflow, I_u = Underground inflow, P_l = Lake precipitation, Q_s = Surface outflow, Q_u = Underground outflow and E_l = Lake evaporation.

The water budget of lakes is determined by the conditions of humidification in the area where the lakes and drainage basins are located. In humid areas, precipitation is higher than evaporation from water surface, and in arid regions, evaporation from water surface is higher than precipitation. An intermediate zone may also exist where both humid and arid zones may alternate. Each zone is characterised by its own structure and dynamics. Quantitatively, water budget of lakes can be computed by using conventional techniques, the drawbacks of which can be overcome using isotopic techniques. The conventional method of computing water budget is through parameters like surface inflow, lake precipitation, lake evaporation, surface outflow, subsurface outflow and inflow and storage changes. Water budget of lakes can also be estimated using environmental isotopes and artificial isotopes.

4.13.1 Sedimentation

The lake bottom is covered by mud due to sedimentation of particles in the water, which comprise inorganic silt derived from surrounding lands and organic particles of dead plants and animals sinking down the water column. These act as food for the bottom dwelling animals and source chemicals to the water. In many lakes, bottom sediments are principal sites of decomposition. Sediments with high organic content contain large numbers of bacteria, which break them into smaller inorganic molecules. These bacteria require oxygen and when the sediment becomes totally devoid of oxygen, they start to adsorb it from the overlying water. If the lake is stratified, then there is no mixing of hypolimnion and hence no replenishment of oxygen to the lower layers. This leads to anoxic conditions and death of animals.

If the lake is large and regularly has a deoxygenated hypolimnion, only those animals able to endure the absence of oxygen will survive. When the sediments are devoid of oxygen, some processes of decomposition cannot occur, but other chemical reactions that occur only in the absence of oxygen take over. Anaerobic bacteria can operate without oxygen and break down large quantities of organic matter in the mud to produce methane gas (made up of carbon and hydrogen CH_4). The released methane is used as energy by other bacteria, which are aerobic and can live only in oxygenated conditions. Other anaerobic bacteria can use sulphates and the by-product of their activity is hydrogen sulphide. Many lakes have a layer of deoxygenated water at the bottom during summer, above which a layer of oxygenated water is present. The balance between these two layers and between the intensity of aerobic and anaerobic activities is essential to the health of the lake ecosystem.

The endangering of lakes due to high rate of sedimentation has caused serious problems, particularly in developing countries. The loss of storage capacity of the lakes due to unwanted sedimentation may have a profound effect on the functioning and objectives of the water system. The solution to this problem requires a comprehensive study of the catchment with regard to the following aspects:

(i) Measurement of erosion effects in the catchment and identification of the sediment sources including those related to particular land use.

(ii) Identification of areas that serve as sediment storage.

(iii) Measurement of temporal changes in sediment storage.

(iv) Determination of sediment rates.

Sedimentation in lakes records the effects of climatological or man-made changes in and around the catchment. The important factors influencing this process in lakes/reservoirs are as follows:

- Nature and size of watersheds surrounding the lake basins; high relief areas contribute coarser particles whereas low-gradient areas contribute finer material.
- Size of the lake/reservoir; controls the wave action within the system and hence affects the sedimentation behaviour.
- Orientation of lake/reservoir relative to dominant wind direction.
- Thermal stratification and density of lake water; particularly important with respect to movement of fine particles.
- Climatic factors such as precipitation, evaporation, etc.

Rivers and streams provide pathways for sediment movement to lakes/reservoirs. The nature of sediment transported ranges from pebbles and coarse sands to fine sands, silts, clays (suspended/wash load), depending on a number of variables such as water discharge, bed slope, velocity, turbulence, etc. Although in most lakes, sedimentation is principally from inorganic material, in many shallow lakes (in areas of low relief, where rivers carry small loads), the indigenous phytoplankton populations may contribute significant quantities of particulate matter to the lakebed.

The most conventional technique and accurate determination of sediment load being carried to a lake by streams is to measure the flow rate and sediment concentration of the inflowing waters just upstream of the lake. The other conventional methods involve periodic bathymetric surveys of the lake. Environmental isotopes can also be used for this purpose. Increased rate of sedimentation can attenuate the light impinging on the surface of water.

4.13.2 Light

Light plays an important role in lake ecology and determines the potential rate of photosynthesis, which supplies dissolved oxygen and food in the water. Solar radiation is the major source of heat and determines the wind patterns in the lake basin and water movement. Nearly all energy that controls the metabolism of lakes and streams is derived directly from the solar energy utilised in photosynthesis. That energy stored in photosynthetically formed organic matter is both synthesised within the lake or stream (autochthonous) or within drainage basin and brought to the lake or stream in various forms (allochthonous). Utilisation of this energy received within the water or imported from the drainage basin and factors that influence the efficiency of conversion of solar energy to potential chemical energy are fundamental to lake productivity. Light intensity varies with seasons and depth. Deeper the light penetration, higher is the rate of photosynthesis. Photosynthetic organisms include phytoplankton (algae), attached algae (periphyton) and vascular aquatic plants (macrophytes). The rate at which the penetration of light decreases with depth depends on the amount of suspended particles in water. The level of light determines the maximum depth at which algae and macrophytes can grow, which is determined to be the point where available light is reduced to 0.5% - 1% of the amount available on the surface. In a clear lake, the euphotic zone may extend down to 20 m or more but in most of the present day lakes, it is often about 3-5 m or as less as 0.5 m deep. Significant changes in lake transparency are the result of anthropogenic activities in the watershed.

4.13.3 Light and Depth

When light passes down through a column of water, the amount of light available at any particular depth decreases exponentially, as most of the light is absorbed in the surface waters. The depth of light penetration depends very much on the colour and turbidity of the water.

Thus, if there are 100 units of light energy at the surface, they will decrease depending on the water quality and not in a simple linear manner (like 90 at 1m, 80 at 2m and so on).

Nutrient cycling, distribution of dissolved gases and biota, and behavioural adaptations of organisms are all markedly influenced by the thermal structure and stratification patterns. Therefore the optical properties of lakes and reservoirs are important regulatory parameters in the physiology and behaviour of aquatic organisms.

4.13.4 Density

Much of the radiation entering the lake, especially those with longer wavelengths, is absorbed near the surface and transformed into heat. Water differs from other compounds as it is less dense as solid than as liquid (most dense at 4°C and less dense at both higher and lower temperatures). When the surface water of the lake warms up, heat takes a long time to penetrate down through the water. During summer, temperature differences between the upper and lower layers become more distinct. Deep lakes generally become stratified into three identifiable layers known as epilimnion, metalimnion and hypolimnion.

The epilimnion is the warm upper layer, which is well mixed. Below the epilimnion is the metalimnion or thermocline region, a layer of water in which the temperature declines rapidly with depth. The hypolimnion is the bottom layer of cold water. Change in density at the metalimnion acts as a physical barrier preventing the mixing of epilimnion and hypolimnion. Thermocline is used synonymously with metalimnion, but is actually the plane or surface of maximum rate of decrease of temperature with respect to depth. Thus, thermocline is the point of maximum temperature change within the metalimnion.

There are few lakes, which are so deep and warm that they are permanently stratified, and many shallow lakes that are never stratified for more than few hours. But most lakes undergo seasonal stratification and mixing, which is a major determinant of their ecology. Thermal stratification and heat exchange depend on solar radiation and wind. In addition to climate, the size and wind exposure of a lake and water inflow are the major factors that determine the type of circulation. Lakes can be categorised according to the degree and frequency of their circulation patterns.

Circulation pattern in lakes:

a. Amictic: never mix as they are permanently frozen.

b. Meromictic: mix only partially, the deeper layers never mix either because of high water density caused by dissolved substances or because the lake is protected from wind effects.

c. Holomictic: mix completely.

d. Oligomictic: do not mix every year as they are large and have higher heat storing capacity, the mixing depending on specific climatic conditions.

e. Monomictic: mix only once each year, either in winter or summer.

f. Dimictic: mix twice a year and are the most common lakes in temperate latitudes.

g. Polymictic: mixing frequently, they are shallow tropical lakes with great wind exposure.

4.13.5 Lake Chemistry

The chemical and biological characteristics of the lake depend on the following:

(i) Formation

(ii) Basin size and shape

(iii) Topography and chemistry

(iv) Regional climate

(v) Biological community

(vi) Anthropogenic activity

Lakes are extremely variable in their physical, chemical and biological characteristics. Physically they vary in terms of the level of light, temperature and water currents. Chemically they vary in nutrients, major ions and contaminants, and biologically in terms of biomass, population numbers and growth.

The chemical composition of a lake is a function of its climate, hydrology and basin geology. Each lake has three major anions and four major cations, which are in ionic balance. Ion balance means the sum of negative ions equals the sum of positive ions when expressed in equivalents. These ions are expressed as mg/L (ppm).

Anthropogenic activities have profound influence on lake chemistry. Excessive landscape disturbance causes higher rates of leaching and erosion by removing vegetative cover, exposing soil and increasing water runoff velocity. Lakes with high concen-trations of calcium and magnesium are called hard water lakes, while those with low concentrations of these ions are called soft water lakes.

H (hydrogen ion) contributes to the acidity of the lake water; while mercury (Hg) another significant pollutant, affecting the aquatic ecosystems, can bioaccumulate in the aquatic food webs, contaminating fish and causing threat to humans and wildlife. Anions and cations carry electrical charges, and the water containing them can conduct electricity and the measurement of the total quantity of charged particles is called as conductivity or TDS (total dissolved solids). TDS concentration and the relative ratios of different ions influence the type of organisms that can best survive in the lake and also affect the chemical reactions occurring in the water.

All the ions listed in the above table are vital to the health of the living organisms in the lake. For eg., calcium is essential for all the cell processes of plants and animals and serves as a structural component for the shells of invertebrate animals like molluscs. Magnesium is important for photosynthesis. Iron is a major component of the red blood pigment haemoglobin (found in all invertebrates) and also for certain cell processes. Silica is essential for the growth of diatoms as their outer cast (frustule) is entirely made up of silica.

4.14 DISSOLVED OXYGEN

Most aquatic organisms require oxygen for their metabolic activities. The supply of oxygen in water comes from the exchange with the atmosphere or from photosynthesis in flora and cyanobacteria (blue green algae). Photosynthesis produces organic matter and releases oxygen, whereas aerobic respiration consumes organic matter and uses oxygen. Oxygen production (photo synthesis) predominates in the light whereas oxygen consumption takes place in the dark. Based on these processes lakes can be divided into two zones namely lighted trophogenic zone (where organic matter is synthesised and oxygen generated) and tropholytic zone (leads to decomposition of organic matter and lowering of oxygen level).

The trophogenic zone often corresponds to the epilimnion. The organic matter produced during photosynthesis sinks to the bottom but the oxygen remains in the surface layers (epilimnion). Hence, there is a sharp delineation between oxygen production and consumption. The entire water mass is oxygenated only during periods of circulation, which occur when the surface waters are disturbed by wind. During stagnation periods (absence of wind), only the epilimnion is oxygenated. The oxygen required for the decomposition of organic matter that sinks to the bottom must therefore come from the supply obtained during periods of circulation. Thus the oxygen levels in the hypolimnion are constantly depleted.

The critical determinants of the oxygen balance in a lake are morphometry and productivity. Among the lakes having same productivity, those with greater depth will decompose organic matter with little changes in the dissolved oxygen while shallow lakes lose oxygen completely. The partially decomposed organic matter settles as sediments in the bottom. The layer of water that is in contact with the sediments tends to lose oxygen as decomposition process continues in the sediment. Lower dissolved oxygen in the water overlying the sediments can aggravate water-quality deterioration because lowered DO levels (below 1 mg/L) initiate chemical reactions at the sediment-water interface, releasing phosphorous from the sediments into the water. When the lake mixes due to wind during summer, phosphorous and ammonium that is built up in the sediment triggers algal growth.

The vertical change in oxygen concentration is dependent on the pattern of circulation. In holomictic lakes, water is saturated with oxygen from the top to the bottom. Such oxygen profiles are called as orthrograde curves. In productive lakes where oxygen levels drop to zero in the hypolimnion, the oxygen curve is called as clinograde curve. Aquatic life in lakes has evolved many adaptations for the range of oxygen conditions in the lake. Many organisms have life cycles adapted to the predictable changes in the oxygen conditions such as obligate anaerobic microorganisms, which can live only in regions devoid of oxygen.

Determination of DO concentrations is a fundamental part of water quality assessment since dissolved oxygen is involved in or influences all the chemical and biological reactions in water. Concentrations below 5 mg/L may adversely affect the functioning and survival of biological communities and below 2 mg/L may lead to fish mortality. The measurement of DO can be used to indicate the degree of pollution by organic matter, destruction of organic matter and level of self purification of the water.

pH

The pH is a measure of the acid-base balance of a solution and is defined as the negative of the logarithm to the base 10 of the hydrogen ion concentration. The pH scale runs from 0 to 14 (i.e., very acidic to very alkaline) with pH 7 representing a neutral solution. At a given temperature pH indicates the acidic or basic character of the solution and is controlled by chemical and biochemical processes in the solution. Acid-base balance of a waterbody can be affected by inflow of industrial effluents, domestic sewage and atmospheric deposition of acid-forming substances. Diurnal variations in pH can take place due to photosynthesis and respiration cycles of algae in eutrophic waters. Acidity of water is controlled by strong mineral acids, weak acids such as carbonic, humic and fulvic acids, and hydrolysing salts of metals such as iron and aluminium.

Vertical difference in biological activity in the lake may lead to vertical changes in pH. The three main processes that affect lake pH are photosynthesis, respiration and nitrogen assimilation. The effect of photosynthesis and respiration depends on carbonate-bicarbonate-carbon dioxide equilibrium. The simplified formulae for photosynthesis using carbon dioxide or bicarbonate are]

$$6CO_2 + 6\,H_2O \rightarrow C_6H_{12}O_6 + 6O_2$$

$$6\,HCO_3 + 6\,H \rightarrow C_6H_{12}O_6 + 6O_2$$

From the above equations it is obvious that no protons are used when carbon-di-oxide is taken up during photosynthesis, whereas one proton per carbon atom is used when bicarbonate ions are used for the same. The reverse happens during respiration. When pH is less than 6.3 and only carbon dioxide is present, photosynthesis and respiration have no effect on the pH. At higher pH values, when other forms of inorganic carbon are available, photosynthesis and respiration alter the uptake and release of protons. This affects the alkalinity or acid-neutralising capacity of the water.

4.14.1 Nutrients

Nutrients are the basic requirements of plants for their growth along with water and sunlight. Aquatic plants and algae respond to even small changes in the amount of nutrients present in the water. Hence it is necessary to estimate the concentrations of nutrients in the lake water and inflows to the lake. The identification of the watershed areas and land use activities that contribute to these nutrients in the lake water is essential. The two most important nutrients contributing to anthropogenic or cultural eutrophication are nitrogen and phosphorous. Both these nutrients are present in sufficient concentrations in fresh waters to maintain a healthy ecosystem, but anthropogenic activities may alter their concen-trations contributing to algal blooms.

Phosphorous and nitrogen enter the lake not only as inorganic ions but also as inorganic polymers, organic compounds, living micro organisms and detritus. Only a few of these forms are readily available for plant and algal growth. A nutrient-poor lake may have only about 1mg/l of phosphorous or 50 mg/L of nitrogen, while the most fertile lake may have up to a milligram of phosphorous or 20-30 mg/L of nitrogen.

4.15 NITROGEN COMPOUNDS

Plants and microorganisms convert inorganic nitrogen to organic nitrogen. The inorganic compounds include nitrite, nitrate, ammonium ions and molecular nitrogen. These undergo biological and non-biological transformations in the environment. The major non-biological transformations include sorption (absorption and adsorption), voltalisation and sedimentation. Biological trans-formations include: a) assimilation of inorganic ions (ammonia and nitrate) by plants and microorganisms to form organic nitrogen (amino acids); b) reduction of nitrogen gas to organic nitrogen and ammonia by microorganisms; c) oxidation of ammonia to nitrite and nitrate (nitrification); d) conversion of organic compounds to ammonia during the decomposition of organic matter; e) bacterial reduction of nitrate to nitrous oxide and molecular nitrogen under anoxic conditions (denitrification).

In lakes where the concentration of nitrogen compounds is extremely low, plants can take up additional inorganic nitrogen immediately. The discharge of sewage into the water body causes large growth of algae (if discharged directly into the lake or river). Raw sewage normally contains complex organic compounds of nitrogen that is decomposed by bacteria before they are used by plants. Eventually bacteria use up all oxygen present in the water, which leads to the death of aquatic plants and animals. Along with sewage, sometimes, fertilisers are washed off from the surrounding agricultural land, which adds to eutrophication (the process of enrichment of lakes due to addition of nutrients).

4.15.1 Ammonia

Ammonia occurs naturally in water due to the breakdown of nitrogenous organic and inorganic matter in soil and water, excretion by biota, reduction of nitrogen gas in water by micro-organisms and gas-exchange in the atmosphere. At certain pH levels, high concentrations of ammonia are toxic to aquatic life and are detrimental to the ecological balance of water bodies. In aqueous solution, un-ionised ammonia exists in equilibrium with ammonium ion. Total ammonia is the sum of these two forms, also forming complexes with several metal ions, and may be adsorbed into colloidal particles, and suspended and settled sediments. The concentrations of un-ionised ammonia depend on pH, temperature and total ammonia concentration. Unpolluted waters may contain small amounts of ammonia and ammonia compounds (less than 0.1mg/L as nitrogen), but may reach 2-3 mg/L N. Higher concentrations are an indication of organic pollution due to sewage, industrial waste and fertiliser run-off. Natural seasonal fluctuations also occur as a result of death and decay of aquatic organisms, particularly phytoplankton and bacteria in nutritionally rich waters.

4.15.2 Nitrate and Nitrite

Nitrate ions are commonly found in natural waters. It may be reduced to nitrite by denitrification process (usually under anaerobic conditions). The nitrite ions rapidly oxidise to nitrate, which is an essential constituent of aquatic plants, although seasonal fluctuations can occur due to plant growth and decay. The natural concentration of 0.3 mg/L of nitrate may be enhanced by fertilisers in the runoff, and industrial and municipal wastewaters (to 5 mg/L). In lakes, concentration of nitrate in excess of 0.2 mg/L of nitrate nitrogen stimulates algal growth leading to eutrophication. Nitrite concentrations in freshwaters are usually low (0.001 mg/L). High nitrite concentrations in water are indicative of poor microbiological quality of water. Determination of nitrite and nitrate in surface waters gives a general indication of the nutrient status and level of organic pollution. As the determination of nitrate is difficult due to interferences from other substances in the water, the precise choice of method may vary according to the expected concentration of nitrogen as nitrate.

4.15.3 Organic Nitrogen

Organic nitrogen consists of protein substances and the product of their transformations. It is subject to seasonal fluctuations of the biological community, formed in water by phytoplankton and bacteria, and recycled within the food chain.

4.15.4 Phosphorous

Phosphorous is an essential nutrient for living organisms and exists in water bodies as dissolved and particulate matter. In natural waters, it occurs mostly as dissolved orthophosphates and polyphosphates and organically bound phosphates. Changes between these forms occur continuously due to the degradation and synthesis of organically bound forms and oxidised inorganic forms. Phosphorous is rarely found in high concentrations in freshwaters and ranges from 0.005 to 0.020 mg/L. High concentrations of phosphates can indicate the presence of pollution and are responsible for eutrophic conditions. Carbon and nitrogen are abundant in natural waters and additionally available from gases in the atmosphere. In majority of lakes, availability of phosphorous is the limiting factor, which controls the rate at which plants grow and therefore the productivity of the whole plant community. Phosphorous is much more readily lost from an ecosystem than nitrogen and carbon as it reacts with mud and chemicals in water in ways that make it unavailable for plants. Plants can absorb phosphorous only as dissolved inorganic phosphorous, which is rapidly taken up by algae and macrophytes.

Bacteria in the sediments at the bottom of the lake break down organic content of dead plants and animals and phosphate is released into the water in the spaces between the sediment particles. This process is rapid in sediments devoid of oxygen. In a lake with oxygenated water, even a thin layer of sediment having oxygen will act as a barrier and prevent the release of phosphate from the sediments below (which is deoxygenated). If water becomes deoxygenated (as in summer), phosphate is slowly released from the sediments. In a shallow lake, however, the phosphate may stay locked in the deep layers, but due to the lack of depth the wind mixes the water and releases the phosphate from the sediment. Aquatic plants and algae absorb the released phosphate in the water and their population's increase. This enhances the death and decomposition of more phosphate containing materials in the water, which in turn reduces the oxygen levels and speeds up the release of more phosphate.

4.16 PHOTOSYNTHESIS AND RESPIRATION

All green plants make their own food by the process known as photosynthesis. They trap energy from sunlight with the aid of chlorophyll and with carbon dioxide and water form sugar. The sugars are stored and used as a source of chemical energy and combine with other molecules to form proteins, oils, fats and carbohydrates. The chemical energy stored as carbohydrates are used for respiration. All plants and animals must respire in order to stay alive, so unlike photosynthesis, it continues even in the dark. During respiration, the oxygen is taken in and combined with cell sugars to release energy. Water and carbon dioxide is released as waste products.

4.16.1 Carbon

Carbon atoms link to form complex organic materials, as carbon is the structural basis of all living material on earth. During photosynthesis, energy is used to link carbon compounds forming sugars, starch and cellulose. Plants take up carbon dioxide and fix it in plant material. The amount of carbon dioxide fixed in the plant material is measured as the amount of carbon fixed per volume of the water. The amount of carbon fixed in a given time is an important measure of the productivity of the lake. Plants obtain carbon from the atmosphere in the form of carbon dioxide and animals from plant material. Aquatic plants use carbon dioxide dissolved in the water but the amount depends on the acidity of the water. Above pH 5.0, some of the carbon dioxide molecules form bicarbonate ions and by pH 9 all the molecules are incorporated into bicarbonate plus some carbonate ions. Above pH 9.5 carbonate ions dominate and there is a lack of free carbon dioxide. Majority of living organi-sms also release carbon dioxide when they respire and this includes many of the bacteria involved in decomposition. These bacteria also release carbon from the bodies of dead plants and animals back into the water for use by living plants.

4.17 ORGANIC MATTER

Organic matter in freshwater arises from living material (directly from photosynthesis or indirectly from terrestrial organic matter) and as a constituent of many waste materials and effluents. The total organic matter can be a useful indication of pollution. In surface waters, concentration of total organic carbon is less than 10 mg/L

Chemical Oxygen Demand (COD) is a measure of the oxygen equivalent of the organic matter in a water sample that is susceptible to oxidation by strong chemical oxidant such as dichromate. The concentration of COD in surface water ranges from 20mg/L oxygen or less in unpolluted waters to greater than 200 mg/L (in waters receiving industrial effluents).

Biochemical Oxygen Demand (BOD) is an approximate measure of the biochemically degradable organic matter present in the water sample. It is defined as the amount of oxygen required for aerobic microorganisms in the sample to oxidise the organic matter to a stable inorganic form. Unpolluted water typically has BOD value of 2mg/L, but those receiving effluent may have more than 10mg/L.

Humic and Fulvic acids: Organic matter from flora and fauna makes a major contribution to the natural quality of surface water, the composition of which is extremely diverse. Natural organic matter is not toxic but exerts major influence on biochemical and hydrochemical processes in the water body. Humus is formed by biochemical and chemical decomposition of vegetative residues and microorganism activity. It enters directly from the soil or is a result of biochemical transformations within the lake. Humus is divided into humic and fulvic acids. These concentrations are highly dependant on the physical-geographical conditions and range from 10 to 100 mg of carbon/litre.

Major ions: These include calcium, magnesium, sodium, potassium, chlorides, sulphates, bicarbonates, etc.

Calcium is present in all surface waters as Ca and is usually dissolved from rocks rich in calcium minerals. The cation is abundant in surface water and along with magnesium is responsible for the hardness in water. It is an important constituent in all organisms and is incorporated into the shells of many invertebrates and bones of vertebrates. Calcium concentration in natural water is <15 mg/L.

Magnesium also contributes to hardness in water. It occurs in many of the organometallic compounds and in organic matter since it is an important element for living organisms. Natural concentrations of magnesium vary from 1 to > 100 mg/L.

The presence of carbonates and bicarbonates influences the hardness and alkalinity of water. Inorganic CO_2 arises from atmosphere and biological respiration. The relative amounts of carbonates, bicarbonates and carbonic acid are related to pH. As a result of weathering of rocks combined with the pH range of surface waters (6-8.2), bicarbonate is found as the dominant anion in most surface waters. Carbonate is uncommon in natural surface waters because they rarely exceed pH 9.0.

Most chlorine occurs as chloride in solution. In a clean surface water, chloride concentrations are less than 10 mg/L and sometimes less than 2 mg/L. Higher concentrations can occur due to sewage, industrial effluents and agricultural and road run-off.

Sulphate is naturally present in surface water as SO_4. It is the stable, oxidised form of sulphur that is readily available in water. Industrial discharges can also add significant amounts of sulphate to surface waters. Sulphate can be used as an oxygen source by bacteria that convert it into hydrogen sulphide (H_2S) under anaerobic conditions. Sulphate concentrations in natural waters are usually between 2mg/L to 80 mg/L and exceed 1000 mg/L near industrial discharge.

Sodium is found in the ionic form and in plant and animal matter. All natural waters contain some sodium, as its salts are highly soluble in water. Sewage and industrial effluents increase the presence of sodium, commonly measured where water is used for drinking or agriculture, particularly irrigation. The surface waters, including those receiving wastewater have concentrations well below 50 mg/L.

Potassium is found usually in the ionic form, its concentration is low in natural water (less than 10 mg/L). Potassium salts are highly soluble. It is readily incorporated into mineral structures and accumulated by aquatic biota, as it is an essential nutrient.

Sulphide formation in surface waters is principally through anaerobic, bacterial decay of organic substances in the sediment. Traces of sulphide ion occur in unpolluted bottom sediments due to the decay of vegetation but higher concentration often indicates the presence of sewage or industrial effluents. Due to its high concentration, toxicity and strong odour make the water unsuitable for drinking and other uses.

Fluoride originates from weathering of fluoride containing minerals and enters surface waters from run-off and groundwater. Its concentrations vary from 0.05 to 100 mg/L.

Metals: The metals in freshwater ecosystems include aluminium, cadmium, chromium, copper, iron, mercury, zinc, nickel, manganese and lead. The ability of a water body to support aquatic life as well as its suitability to other uses depends on the concentration of metals. Some metals are important for physiological functions of living tissue and regulate many biochemical processes. The same metals, discharged at increased concentrations in sewage and industrial effluents, can have severe toxicological effects on humans and aquatic ecosystems. Due to lack of proper elimination measures, metals move from one trophic level to the other bioaccumulating, often with detrimental effects. The toxicity of metals depends on the degree of oxidation of a given metal ion together with the forms in which it occurs. Metals in natural waters can exist in truly dissolved, colloidal and suspended forms. The concentrations of different metals in water vary over a wide range (0.1 - 0.001 mg/L) and can rise to higher concentrations due to anthropogenic activities

4.18 AQUATIC TOXICOLOGY

Aquatic toxicology is the study of the effects of manufactured chemicals and other anthropogenic and natural materials and activities on aquatic organisms at various levels of organization, from subcellular through individual organisms to communities and ecosystems. Aquatic toxicology is a multidisciplinary field which integrates toxicology, aquatic ecology and aquatic chemistry.

This field of study includes freshwater, marine water and sediment environments. Common tests include standardized acute and chronic toxicity tests lasting 24–96 hours (acute test) to 7 days or more (chronic tests). These tests measure endpoints such as survival, growth, reproduction, that are measured at each concentration in a gradient, along with a control test. Typically using selected organisms with ecologically relevant sensitivity to toxicants and a well-established literature background. These organisms can be easily acquired or cultured in lab and are easy to handle.

4.18.1 History of Toxicology

While basic research in toxicology began in multiple countries in the 1800s, it was not until around the 1930s that the use of acute toxicity testing, especially on fish, was established. Over the next two decades, the effects of chemicals and wastes on non-human species became more of a public issue and the era of the pickle-far bioassays began as efforts increased to standardize toxicity testing techniques. In the United States of America, the passage of the Federal Water Pollution Control Act of 1947 marked the first comprehensive legislation for the control of water pollution and was followed by the Federal Water Pollution Control Act in 1956. In 1962, public and governmental interests were renewed, in large part due to the publication of Rachel Carson's Silent Spring, and thee years later the Water Quality Act was passed which directed states to develop water quality standards. Public awareness, as well as scientific and governmental concern, continued to grow throughout the 1970s and by the end of the decade research had expanded to include hazard evaluation and risk analysis. In the subsequent decades, aquatic toxicology has continued to expand and internationalize so that there is now a strong application of toxicity testing for environmental protection.

4.18.2 Aquatic Toxicity Tests

Aquatic toxicology tests (assays): toxicity tests are used to provide qualitative and quantitative data on adverse (deleterious) effects on organisms from a toxicant. Toxicity tests can be used to assess the potential for damage to an aquatic environment and provide a database that can be used to assess the risk associated within a situation for a specific toxicant. Aquatic toxicology tests can be performed in the field or in the laboratory. Field experiments generally refer to multiple species exposure and laboratory experiments generally refer to single species exposure. A dose response relationship is most commonly used with a sigmoidal curve to quantify the toxic effects at a selected end-point or criteria for effect (i.e. death or other adverse effect to the organism). Concentration is on the x-axis and percent inhibition or response is on the y-axis.

The criteria for effects, or endpoints tested for, can include lethal and sublethal effects.

There are different types of toxicity tests that can be performed on various test species. Different species differ in their susceptibility to chemicals, most likely due to differences in accessibility, metabolic rate, excretion rate, genetic factors, dieteary factors, age, sex, health and stress level of the organism. Common standard test species are the fathead minnow(Pimephales promelas), daphnids *(Daphnia magna, D. pulex, D. pulicaria, Ceriodaphnia dubia), midge (Chironomus tentans, C. ruparius),* rainbow trout (Oncorhynchus mykiss), sheepshead minnow *(Cyprinodon variegatu),* mysids*(Mysidopsis),* oyster *(Crassotreas),* scud (Hyalalla Azteca), grass shrimp *(Palaemonetes pugio),* mussels (Mytilus). As defined by ASTM, these species are routinely selected on the basis of availability, commercial, recreational, and ecological importance, past successful use, and regulatory use.

A variety of acceptable standardized test methods have been published. Some of the more widely accepted agencies to publish methods are: the American Public Health Association, U.S. Environmental Protection Agency, American Society for Testing and Materials, International Organization for Standardization, Environment Canada, and Organization for Economic Cooperation and Development. Standardized tests offer the ability to compare results between laboratories.

There are many kinds of toxicity tests widely accepted in the scientific literature and regulatory agencies. The type of test used depends on many factors: Specific regulatory agency conducting the test, resources available, physical and chemical characteristics of the environment, type of toxicant, test species available, laboratory vs. field testing, end-point selection, and time and resources available to conduct the assays are some of the most common influencing factors on test design.

4.19 EXPOSURE SYSTEMS

Exposure systems are four general techniques the controls and test organisms are exposed to the dealing with treated and diluted water or the test solutions.

Static- a static test exposes the organism in still water. The toxicant is added to the water in order to obtain the correct concentrations to be tested. The control and test organisms are placed in the test solutions and the water is not changed for the entirety of the test.

Recirculation- a recirculation test exposes the organism to the toxicant in a similar manner as the static test, except that the test solutions are pumped through an apparatus (i.e. filter) to maintain water quality, but not reduce the concentration of the toxicant in the water. The water is circulated through the test chamber continuously, similar to an aerated fish tank. This type of test is expensive and it is unclear whether or not the filter or aerator has an effect on the toxicant.

Renewal- a renewal test also exposes the organism to the toxicant in a similar manner as the static test because it is in still water. However, in a renewal test the test solution is renewed periodically (constant intervals) by transferring the organism to a fresh test chamber with the same concentration of toxicant.

Flow-through- a flow through test exposes the organism to the toxicant with a flow into the test chambers and then out of the test chambers. The once-through flow can either be intermittent or continuous. A stock solution of the correct concentrations of contaminant must be previously prepared. Metering pumps or diluters will control the flow and the volume of the test solution, and the proper proportions of water and contaminant will be mixed.

4.20 TYPES OF TESTS

Acute tests are short-term exposure tests (hours or days) and generally use lethality as an endpoint. In acute exposures, organisms come into contact with higher doses of the toxicant in a single event or in multiple events over a short period of time and usually produce immediate effects, depending on absorption time of the toxicant. These tests are generally conducted on organisms during a specific time period of the organism's life cycle, and are considered partial life cycle tests. Acute tests are not valid if mortality in the control sample is greater than 10%. Results are reported in EC50, or concentration that will affect fifty percent of the sample size.

Chronic tests are long-term tests (weeks, months years), relative to the test organism's life span (>10% of life span), and generally use sub-lethal endpoints. In chronic exposures, organisms come into contact with low, continuous doses of a toxicant. Chronic exposures may induce effects to acute exposure, but can also result in effects that develop slowly. Chronic tests are generally considered full life cycle tests and cover an entire generation time or reproductive life cycle ("egg to egg"). Chronic tests are not considered valid if mortality in the control sample is greater than 20%. These results are generally reported in NOECs (No observed effects level) and LOECs (Lowest observed effects level).

Early life stage tests are considered as subchronic exposures that are less than a complete reproductive life cycle and include exposure during early, sensitive life stages of an organism. These exposures are also called critical life stage, embryo-larval, or egg-fry tests. Early life stage tests are not considered valid if mortality in the control sample is greater than 30%.

Short-term sublethal tests are used to evaluate the toxicity of effluents to aquatic organisms. These methods are developed by the EPA, and only focus on the most sensitive life stages. Endpoints for these test include changes in growth, reproduction and survival. NOECs, LOECs and EC50s are reported in these tests.

Bioaccumulation tests are toxicity tests that can be used for hydrophobic chemicals that may accumulated in the fatty tissue of aquatic organisms. Toxicants with low solubilities in water generally can be stored in the fatty tissue due to the high lipid content in this tissue. The storage of these toxicants within the organism may lead to cumulative toxicity. Bioaccumulation tests use bioconcentration factors (BCF) to predict concentrations of hydrophobic contaminants in organisms. The BCF is the ratio of the average concentration of test chemical accumulated in the tissue of the test organism (under steady state conditions) to the average measured concentration in the water.

Freshwater tests and saltwater tests have different standard methods, especially as set by the regulatory agencies. However, these tests generally include a control (negative and/

or positive), a geometric dilution series or other appropriate logarithmic dilution series, test chambers and equal numbers of replicates, and a test organism. Exact exposure time and test duration will depend on type of test (acute vs. chronic) and organism type. Temperature, water quality parameters and light will depend on regulator requirements and organism type.

Effluent toxicity tests are tests conducted under the Clean Water Act, National Pollutant Discharge Elimination System (NPDES) permit program and are used by dischargers of contaminated effluent to monitor the quality of effluent into receiving waters. Acute Effluent Toxicity Tests are used to monitor the quality of industrial effluent monthly using acute toxicity tests. Effluent is used to perform static-acute multi concentration toxicity tests with Ceriodaphnia dubia and Pimephales promelas. The test organisms are exposed for 48 hours under static conditions with five concentrations of the effluent. Short-term Chronic Effluent Toxicity Tests are used to monitor the quality of municipal wastewater treatment plants effluent quarterly using short-term chronic toxicity tests. The goal of this test is to ensure that the wastewater is not chronically toxic. The major deviation in the short-term chronic effluent toxicity tests and the acute effluent toxicity tests is that the short-term chronic test lasts for seven days and the acute test lasts for 48 hours.

Sediment Tests: At some point most chemicals originating from both anthropogenic and natural sources accumulate in sediment. For this reason, sediment toxicity can play a major role in the adverse biological effects seen in aquatic organisms, especially those inhabiting benthic habitats. A recommended approach for sediment testing is to apply the Sediment Quality Triad (SQT) which involves simultaneously examining sediment chemistry, toxicity, and field alterations so that more complete information can be gathered. Collection, handling, and storage of sediment can have an effect on bioavailability and for this reason standard methods have been developed to suit this purpose.

4.21 TOXICOLOGICAL EFFECTS

Toxicity can be broken down into two broad categories of direct and indirect toxicity. Direct toxicity results from a toxicant acting at the site of action in or on the organism. Indirect toxicity occurs with a change in the physical, chemical, or biological environment.

Lethality is most common effect used in toxicology and used as an endpoint for acute toxicity tests. While conducting chronic toxicity tests sublethal effects are endpoints that are looked at. These endpoints include behavioral, physiological, biochemical, histological changes.

There are a number of effects that occur when an organism is simultaneously exposed to two or more toxicants. These effects include additive effects, synergistic effects, potentiation effects, and antagonistic effects. An additive effect occurs when combined effect is equal to a combination or sum of the individual effects. A synergistic effect occurs when the combination of effects is much greater than the two individual effects added together. Potentiation is an effect that occurs when an individual chemical has no effect is added to a toxicant and the combination has a greater effect than just the toxicant alone. Finally, an antagonistic effect occurs when a combination of chemicals has less of an effect than the sum of their individual effects.

4.22 IMPORTANT AQUATIC TOXICOLOGY RESOURCES

American Society for Testing and Materials (ASTM International)–A consensus organization, representing 135 countries, that develops and delivers international voluntary standard methods for aquatic toxicity testing.

Standard Methods for the Examination of Water and Wastewater–A compilation of techniques for the examination of water, jointly published by the American Public Health Association (APHA), the American Water Works Association (AWWA), and the Water Pollution Control Federation (WPCF).

Ecotox – A database maintained by the U.S. Environmental Protection Agency (EPA) that offers single chemical toxicity information for both aquatic and terrestrial purposes.

Society of Environmental Toxicology and Chemistry (SETAC) – A nonprofit, worldwide society working to promote scientific research to further our understanding of environmental stressors, environmental education, and the use of science in environmental policy.

United States Environmental Protection Agency (EPA) – A federal agency working to protect human and environmental health. Among many other functions, the U.S. EPA produces guidance manuals outlining aquatic toxicity test procedures.

Organisation for Economic Co-operation and Development (OECD) – A forum for governments to work together to promote policies for the betterment of people's social and economic well-being around the world. One way in which they accomplish this is through the development of aquatic toxicity test guidelines.

Environment Canada (EC) – A diverse organization working to protect Canada's water resources and the natural environment through the coordination of environmental policies and programs with the federal government.

4.23 TERMINOLOGY OF AQUATIC TOXICOLOGY

Median Lethal Concentration (LC50) – The chemical concentration that is expected to kill 50% of a group of organisms.

Median Effective Concentration (EC50) – The chemical concentration that is expected to have one or more specified effects in 50% of a group of organisms.

Critical Body Residue (CBR) – An approach that routinely examines whole-body chemical concentrations of an exposed organism that is associated with an adverse biological response.

Baseline toxicity – Refers to narcosis which is a depression in biological activity due to toxicants being present in the organism.

Biomagnification – The process by which the concentration of a chemical in the tissues of an organism increases as it passes through several levels in the food web.

Lowest Observed Effect Concentration (LOEC) – The lowest test concentration that has a statistically significant effect over a specified exposure time.

No Observed Effect Concentration (NOEC) – The highest test concentration for which no effect is observed relative to a control over a specified exposure time.

Maximum Acceptable Toxicant Concentrations (MATC) – An estimated value that represents the highest "no-effect" concentration of a specific substance within the range including the NOEC and LOEC.

Application Factor (AF) – An empirically derived "safe" concentration of a chemical.

Biomonitoring – The consistent use of living organisms to analyze environmental changes over time.

Effluent – Liquid, industrial discharge that usually contain varying chemical toxicants.

Quantitative Structure-Activity Relationship (QSAR) – A method of modeling the relationship between biological activity and the structure of organic chemicals.

Mode of Action – A set of common behavioral or physiological signs that represent a type of adverse response.

Mechanism of Action – The detailed events that take place at the molecular level during an adverse biological response.

KOW – The octanol-water partition coefficient which represents the ratio of the concentration of octanol to the concentration of chemical in the water.

Bioconcentration Factor (BCF) – The ratio of the average chemical concentration in the tissues of the organism under steady-state conditions to the average chemical concentration measured in the water to which the organisms are exposed.

All terms were derived from Rand.

4.24 SIGNIFICANCE TO REGULATORY WORLD

In the United States aquatic toxicology plays an important role in the NPDES wastewater permit program. In addition to analytical testing for known pollutants, aquatic, whole effluent toxicity tests have been standardized and are performed routinely as a tool for evaluating the potential harmful effects of effluents discharged into surface waters.

For the Clean Water Act under United States Environmental Protection Agency there are water quality criteria and water quality standards derived from aquatic toxicity tests.

4.25 SEDIMENT QUALITY GUIDELINES

While sediment quality guidelines are not meant for regulation, they provide a way to rank and compare sediment quality developed by National Oceanic and Atmospheric Administration(NOAA). These sediment quality guidelines are summarized in NOAA's Screening Quick Reference Tables (SQuiRT) for many different chemicals.

4.26 MARINE POLLUTION

Marine pollution occurs when harmful, or potentially harmful, effects result from the entry into the ocean of chemicals, particles, industrial, agricultural and residential waste, noise, or the spread of invasive organisms. Most sources of marine pollution are land based. The pollution often comes from nonpoint sources such as agriculturalrunoff, wind-blown debris and dust. Nutrient pollution, a form of water pollution, refers to contamination by excessive inputs of nutrients. It is a primary cause of eutrophication of surface waters, in which excess nutrients, usually nitrogen or phosphorus, stimulate algae growth. Many potentially toxic chemicals adhere to tiny particles which are then taken up by plankton and benthos animals, most of which are either deposit or filter feeders. In this way, the toxins are concentrated upward within ocean food chains. Many particles combine chemically in a manner highly depletive of oxygen, causing estuaries to becomeanoxic.

When pesticides are incorporated into the marine ecosystem, they quickly become absorbed into marine food webs. Once in the food webs, these pesticides can cause mutations, as well as diseases, which can be harmful to humans as well as the entire food web.

Toxic metals can also be introduced into marine food webs. These can cause a change to tissue matter, biochemistry, behaviour, reproduction, and suppress growth in marine life. Also, many animal feeds have a high fish meal or fish hydrolysate content. In this way, marine toxins can be transferred to land animals, and appear later in meat and dairy products.

4.26.1 History of Marine Pollution

Although marine pollution has a long history, significant international laws to counter it were only enacted in the twentieth century. Marine pollution was a concern during several United Nations Conferences on the Law of the Seabeginning in the 1950s. Most scientists believed that the oceans were so vast that they had unlimited ability to dilute, and thus render pollution harmless.

In the late 1950s and early 1960s, there were several contro-versies about dumping radioactive waste off the coasts of the United States by companies licensed by the Atomic Energy Commission, into the Irish Sea from the British reprocessing facility at Windscale, and into the Mediterranean Sea by the French Commissariat à l'Energie Atomique. After the Mediterranean Sea controversy, for example, Jacques Cousteau became a worldwide figure in the campaign to stop marine pollution. Marine pollution made further international headlines after the 1967 crash of the oil tanker Torrey Canyon, and after the 1969 Santa Barbara oil spill off the coast of California.

Marine pollution was a major area of discussion during the 1972 United Nations Conference on the Human Environment, held in Stockholm. That year also saw the signing of the Convention on the Prevention of Marine Pollution by Dumping of Wastes and Other Matter, sometimes called the London Convention. The London Convention did not ban marine pollution, but it established black and gray lists for substances to be banned (black) or regulated by national authorities (gray). Cyanide and high-level radioactive waste, for example, were put on the black list. The London Convention applied only to waste dumped from ships, and thus did nothing to regulate waste discharged as liquids from pipelines.

4.26.2 Pathways of Pollution

There are many different ways to categorize, and examine the inputs of pollution into our marine ecosystems. Patin (n.d.) notes that generally there are three main types of inputs of pollution into the ocean: direct discharge of waste into the oceans, runoff into the waters due to rain, and pollutants that are released from the atmosphere.

One common path of entry by contaminants to the sea are rivers. The evaporation of water from oceans exceeds precipitation. The balance is restored by rain over the continents entering rivers and then being returned to the sea. The Hudson inNew York State and the Raritan in New Jersey, which empty at the northern and southern ends of Staten Island, are a source of mercury contamination ofzooplankton (copepods) in the open ocean. The highest concentration in the filter-feeding copepods is not at the mouths of these rivers but 70 miles south, nearer Atlantic City, because water flows close to the coast. It takes a few days before toxins are taken up by the plankton.

Pollution is often classed as point source or nonpoint source pollution. Point source pollution occurs when there is a single, identifiable, and localized source of the pollution. An example is directly discharging sewage and industrial waste into the ocean. Pollution such as this occurs particularly in developing nations. Nonpoint source pollution occurs when the pollution comes from ill-defined and diffuse sources. These can be difficult to regulate. Agricultural runoff and wind blown debris are prime examples.

4.26.3 Direct Discharge

Pollutants enter rivers and the sea directly from urban sewerage and industrial waste discharges, sometimes in the form of hazardous and toxic wastes.

Inland mining for copper, gold. etc., is another source of marine pollution. Most of the pollution is simply soil, which ends up in rivers flowing to the sea. However, some minerals discharged in the course of the mining can cause problems, such ascopper, a common industrial pollutant, which can interfere with the life history and development of coral polyps. Mining has a poor environmental track record. For example, according to the United States Environmental Protection Agency, mining has contaminated portions of the headwaters of over 40% of watersheds in the western continental US. Much of this pollution finishes up in the sea.

4.27 LAND RUNOFF

Surface runoff from farming, as well as urban runoff and runoff from the construction of roads, buildings, ports, channels, and harbours, can carry soil and particles laden with carbon, nitrogen, phosphorus, and minerals. This nutrient-rich water can cause fleshy algae and phytoplankton to thrive in coastal areas; known as algal blooms, which have the potential to create hypoxic conditions by using all available oxygen.

Polluted runoff from roads and highways can be a significant source of water pollution in coastal areas. About 75 percent of the toxic chemicals that flow into Puget Sound are carried by stormwater that runs off paved roads and driveways, rooftops, yards and other developed land.

4.28 SHIP POLLUTION

Ships can pollute waterways and oceans in many ways. Oil spills can have devastating effects. While being toxic to marine life, polycyclic aromatic hydrocarbons (PAHs), found incrude oil, are very difficult to clean up, and last for years in the sediment and marine environment.

Oil spills are probably the most emotive of marine pollution events. However, while a tanker wreck may result in extensive newspaper headlines, much of the oil in the world's seas comes from other smaller sources, such as tankers discharging ballast water from oil tanks used on return ships, leaking pipelines or engine oil disposed of down sewers.

Discharge of cargo residues from bulk carriers can pollute ports, waterways and oceans. In many instances vessels intentionally discharge illegal wastes despite foreign and domestic regulation prohibiting such actions. It has been estimated that container shipslose over 10,000 containers at sea each year (usually during storms). Ships also createnoise pollution that disturbs natural wildlife, and water from ballast tanks can spread harmful algae and other invasive species.

Ballast water taken up at sea and released in port is a major source of unwanted exotic marine life. The invasive freshwater zebra mussels, native to the Black, Caspian and Azov seas, were probably transported to the Great Lakes via ballast water from a transoceanic vessel. Meinesz believes that one of the worst cases of a single invasive species causing harm to an ecosystem can be attributed to a seemingly harmless jellyfish. Mnemiopsis leidyi, a species of comb jellyfish that spread so it now inhabits estuaries in many parts of the world. It was first introduced in 1982, and thought to have been transported to the Black Sea in a ship's ballast water. The population of the jellyfish shot up exponentially and, by 1988, it was wreaking havoc upon the local fishing industry. "The anchovy catch fell from 204,000 tons in 1984 to 200 tons in 1993; sprat from 24,600 tons in 1984 to 12,000 tons in 1993; horse mackerel from 4,000 tons in 1984 to zero in 1993." Now that the jellyfish have exhausted the zooplankton, including fish larvae, their numbers have fallen dramatically, yet they continue to maintain a stranglehold on the ecosystem.

Invasive species can take over once occupied areas, facilitate the spread of new diseases, introduce new genetic material, alter underwater seascapes and jeopardize the ability of native species to obtain food. Invasive species are responsible for about $138 billion annually in lost revenue and management costs in the US alone.

4.29 ATMOSPHERIC POLLUTION

Another pathway of pollution occurs through the atmosphere. Wind blown dust and debris, including plastic bags, are blown seaward from landfills and other areas. Dust from the Sahara moving around the southern periphery of the subtropical ridgemoves into the Caribbean and Florida during the warm season as the ridge builds and moves northward through the subtropical Atlantic. Dust can also be attributed to a global transport from the Gobi and Taklamakan deserts across Korea, Japan, and the Northern Pacific to the Hawaiian Islands. Since 1970, dust outbreaks have worsened due to periods of drought in Africa. There is a large variability in dust transport to the Caribbean and Florida from year to year; however, the flux is greater during positive phases of the North Atlantic Oscillation. The USGS links dust events to a decline in the health of coral reefs across the Caribbean and Florida, primarily since the 1970s.

Climate change is raising ocean temperatures and raising levels of carbon dioxide in the atmosphere. These rising levels of carbon dioxide are acidifying the oceans. This, in turn, is altering aquatic ecosystems and modifying fish distributions, with impacts on the sustainability of fisheries and the livelihoods of the communities that depend on them. Healthy ocean ecosystems are also important for the mitigation of climate change.

4.30 DEEP SEA MINING

Deep sea mining is a relatively new mineral retrieval process that takes place on the ocean floor. Ocean mining sites are usually around large areas of polymetallic nodules or active and extinct hydrothermal vents at about 1,400 - 3,700 m below the ocean's surface. The vents create sulfide deposits, which contain valuable metals such as silver, gold, copper,manganese, cobalt, and zinc. The deposits are mined using either hydraulic pumps or bucket systems that take ore to the surface to be processed. As with all mining operations, deep sea mining raises questions about potential environmental impact on surrounding areas. Environmental advocacy groups such as Greenpeace have argued that seabed mining should not be permitted in most of the world's oceans because of the potential for damage to deepsea ecosystems.

4.30.1 Brief History

In the mid 1960s the prospect of deep-sea mining was brought up by the publication of J. L. Mero's Mineral Resources of the Sea. The book claimed that nearly limitless supplies of cobalt, nickel and other metals could be found throughout the planet's oceans. Mero stated that these metals occurred in deposits of manganese nodules, which appear as lumps of compressed sediment on the sea floor at depths of about 5,000 m. Some nations including France, Germany and theUnited States sent out research vessels in search of nodule deposits. Initial estimates of deep sea mining viability turned out to be much exaggerated. This overestimate, coupled with depressed metal prices, led to the near abandonment of nodule mining by 1982. From the 1960s to 1984 an estimated US $650 million had been spent on the venture, with little to no return.

Over the past decade a new phase of deep-sea mining has begun. Rising demand for precious metals in Japan, China,Korea and India has pushed these countries in search of new sources. Interest has recently shifted toward hydrothermal vents as the source of metals instead of scattered nodules. The trend of transition towards an electricity-based information and transportation infrastructure currently seen in western societies further pushes demands for precious metals. The current revived interest in phosphorus nodule mining at the seafloor stems from phosphor-based artificial fertilizers being of significant importance for world food production. Growing world population pushes the need for artificial fertilizers or greater incorporation of organic systems within agricultural infrastructure.

Currently, the best potential deep sea site, the Solwara 1 Project, has been found in the waters off Papua New Guinea, a high grade copper-gold resource and the world's first Seafloor Massive Sulphide (SMS) resource. The Solwara 1 Project is located at 1600 metres water depth in the Bismarck Sea, New Ireland Province. Using ROV (remotely operated underwater vehicles) technology, Nautilus Minerals Inc. is first company of its kind to announce plans to begin full-scale undersea excavation of mineral deposits. However a dispute with the government of Papua-New Guinea has delayed production from its expected start in early 2013.

4.30.2 Laws and Regulations

The international law–based regulations on deep sea mining are contained in the United Nations Conventions on the Law of the Sea from 1973 to 1982, which came into force in 1994. The convention set up the International Seabed Authority(ISA), which regulates nations' deep sea mining ventures outside each nations' Exclusive Economic Zone (a 200-nautical-mile (370 km) area surrounding coastal nations). The ISA requires nations interested in mining to explore two equal mining sites and turn one over to the ISA, along with a transfer of mining technology over a 10- to 20-year period. This seemed reasonable at the time because it was widely believed that nodule mining would be extremely profitable. However, these strict requirements led some industrialized countries to refuse to sign the initial treaty in 1982.

4.30.3 Resources Mined

The deep sea contains many different resources available for extraction, including silver, gold, copper, manganese, cobalt, and zinc. These raw materials are found in various forms on the sea floor, usually in higher concentrations than terrestrial mines.

4.30.4 Minerals and Related Depths

Diamonds are also mined from the seabed by De Beers and others. Nautilus Minerals Inc and Neptune Minerals are planning to mine the offshore waters of Papua New Guinea and New Zealand.

4.30.5 Extraction Methods

Recent technological advancements have given rise to the use remotely operated vehicles (ROVs) to collect mineral samples from prospective mine sites. Using drills and other cutting tools, the ROVs obtain samples to be analyzed for precious materials. Once a site has been located, a mining ship or station is set up to mine the area. There are two predominant forms of mineral extraction being considered for full scale operations: continuous-line bucket system (CLB) and the hydraulic suction system. The CLB system is the preferred method of nodule collection. It operates much like a conveyor-belt, running from the sea floor to the surface of the ocean where a ship or mining platform extracts the desired minerals, and returns the tailings to the ocean. Hydraulic suction mining lowers a pipe to the seafloor which transfers nodules up to the mining ship. Another pipe from the ship to the seafloor returns the tailings to the area of the mining site.

In recent years, the most promising mining areas have been the Central and Eastern Manus Basin around Papua New Guinea and the crater of Conical Seamount to the east. These locations have shown promising amounts of gold in the area's sulfide deposits (an average of 26 parts per million). The relatively shallow water depth of 1050 m, along with the close proximity of a gold processing plant makes for an excellent mining site.

4.30.6 Environmental Impacts

Because deep sea mining is a relatively new field, the complete consequences of full scale mining operations are unknown. However, some researchers have said they believe that removal of parts of the sea floor will result in disturbances to thebenthic layer, increased toxicity of the water column and sediment plumes from tailings. Removing parts of the sea floor could disturb the habitat of benthic organisms, with unknown long-term effects. Aside from the direct impact of mining the area, some researchers and environmental activists have raised concerns about leakage, spills and corrosion that could alter the mining area's chemical makeup. Among the impacts of deep sea mining, sediment plumes could have the greatest impact. Plumes are caused when the tailings from mining (usually fine particles) are dumped back into the ocean, creating a cloud of particles floating in the water. Two types of plumes occur: near bottom plumes and surface plumes. Near bottom plumes occur when the tailings are pumped back down to the mining site. The floating particles increase the turbidity, or cloudiness, of the water, cloggingfilter-feeding apparatuses used by benthic organisms. Surface plumes cause a more serious problem. Depending on the size of the particles and water currents the plumes could spread over vast areas. The plumes could impactzooplankton and light penetration, in turn affecting the food web of the area.

4.30.7 Current Events and News

The Nautilus Minerals Corporation is currently in talks with the government of Papua New Guinea, attempting to set up and start a deep sea mining project

5

WATER CONSERVATION METHODS

5.1 OBJECTIVES

After studying this chapter you will be able to understand:

- Ancient indian methods of water conservation.
- Water conservation policy as a solution to water scarcity.
- Minimum water network target and design
- Importance of water to plants.
- Water erosion problem in india.

5.2 INTRODUCTION

Water conservation is a big thing, but every little bit helps, so don't think that what you do doesn't matter. A whole lot of people doing a little bit adds up to a whole lot. We must all make changes in our lifestyles that will change the course of our water and its quality. Water conservation needs to be a way of life, not just something we think about once in a while. If we all do our part in conserving Washington's water, we can make a huge difference for the environment.

Water conservation means using our water wisely and caring for it properly. Since each of us depends on water for life, it is our responsibility to learn more about water conservation and how we can help keep our water pure and safe for generations to come. Since we all enjoy the benefits of having pure, clean water, we must help conserve water so that we may continue to enjoy these benefits.

Water conservation is not a job that is just for the technician, soil scientist, hydrologist, forester, wildlife manager, plant scientist, city planner, park manager, farmer, rancher, or mine owner alone. It is a job for the every day person who just likes to have access to the life sustaining resource of water. We must all recognize that water conservation really is our personal responsibility and not just leave it up to other people. We all enjoy the many ways that we use water, so why not do our part in caring for our water?

Our water supply is finite, which means that we do not have an endless supply. We only have the water that we have now. Ninety - seven percent of all the water on the earth is salt water which is not suitable for drinking. Only three percent of all the water is fresh water, and only one percent is available for drinking water. The other two percent is locked in ice caps and glaciers.

With all the people on Earth relying on such a small percentage of all the water on Earth, it only makes sense that we must preserve and conserve our water. We must not pollute our water because it is the only water we will ever have. Some people do not realize the importance of water, and they are continually polluting it. Only about ten percent of waste water is disposed of properly.

We must learn to save water now for the future. The quality of our water is very important. We have the same amount of water now as there was when the earth was created. This is the water we have, and we must preserve its quality. Water is the foundation of food and life. Next to air, water is our most precious resource. We cannot live without water. Saving water helps to preserve our environment. It reduces the energy required to process and deliver water, which helps in reducing pollution and in conserving fuel resources. Saving water now means having water available in the future for recreational purposes, too. Conserving the water we have minimizes the effects of water shortages and helps build a better defense against future drought years. If we save water now, we are helping to ensure a water supply adequate for future generations. Saving water saves money.

We must save water today so that it will be available to us in the future. We need to think of future generations, people who will not have a sufficient supply of water unless we become more concerned with how we use our water today. There are many things that we can do to help preserve Washington's water. These are not things that are hard. All it takes is a little bit of extra effort, and soon it will be second nature. Save the water, its all we've got!

Water Conservation is the most important step in the direction of finding solutions to issues of water and environmental conservation is to change people's attitudes and habits--this includes each one of us. Conserve water because it is the right thing to do. A step to conserve water is the step to secure the future. The most essential among all the natural resources on earth is water. A drop of water is worth more than a sack of gold for the thirsty man. If each one of us make efforts to save water today, it will save us later. Water conservation is the most effective and environmentally sound method to fight global warming. Water conservation is what that can reduce the scarcity of water. It aims to improve the efficiency of use of water, and reduce losses and waste. We can follow some of the simple things that have been listed below and contribute to water conservation.

- Try to do one thing each day that will result in saving water. Don't worry if the savings are minimal¾every drop counts! You can make a difference.
- Remember to use only the amount you actually need.
- Form a group of water-conscious people and encourage your friends and neighbours to be part of this group. Promote water conservation in community newsletters and on bulletin boards. Encourage your friends, neighbours and co-workers to also contribute.
- Encourage your family to keep looking for new ways to conserve water in and around your home.
- Make sure that your home is leak-free. Many homes have leaking pipes that go unnoticed.
- Do not leave the tap running while you are brushing your teeth or soaping your face.
- See that there are no leaks in the toilet tank. You can check this by adding colour to the tank. If there is a leak, colour will appear in the toilet bowl within 30 minutes. (Flush as soon as the test is done, since food colouring may stain the tank.)
- Avoid flushing the toilet unnecessarily. Put a brick or any other device that occupies space to cut down on the amount of water needed for each flush.
- When washing the car, use water from a bucket and not a hosepipe.
- Do not throw away water that has been used for washing vegetables, rice or dals--use it to water plants or to clean the floors, etc.
- You can store water in a variety of ways. A simple method is to place a drum on a raised platform directly under the rainwater collection source. You can also collect water in a bucket during the rainy season.

Our ancient religious texts and epics give a good insight into the water storage and conservation systems that prevailed in those days.

Over the years rising populations, growing industrialization, and expanding agriculture have pushed up the demand for water. Efforts have been made to collect water by building dams and reservoirs and digging wells; some countries have also tried to recycle and desalinate (remove salts) water. Water conservation has become the need of the day. The idea of ground water recharging by harvesting rainwater is gaining importance in many cities.

In the forests, water seeps gently into the ground as vegetation breaks the fall. This groundwater in turn feeds wells, lakes, and rivers. Protecting forests means protecting water 'catchments'. In ancient India, people believed that forests were the 'mothers' of rivers and worshipped the sources of these water bodies.

5.3 MEANING AND SCOPE OF WATER CONSERVATION

Water conservation is a broad term involving all the measures which aim at arresting maximum amount of run-off water, storing or collecting it in safe, artificially raised structures and recycling the stored water again to the field in times of need. The ultimate aim of water conservation is to produce maximum from the limited availability of water, by using it most efficiently. An efficient system of collecting, storing and reusing run-off water is sometimes also referred to as water harvesting.

From the point of dry land agriculture, water harvesting indicates not only storing rain water in irrigation tanks/ponds but also employing mechanical and cultural manipulations of soil to improve its rainfall absorption and moisture retention capacity. However, the measures adopted are location specific.

To decide on the suitable measures to be taken a preliminary inventory of soil water and other resources, a study of the land use pattern, a dialogue with the target group and a study of socio-political, economic and cultural setting are a must. This implies a scientific approach to water conservation. The need for educating, and organizing people to make them accept necessary reforms for effective implementation of integrated water conservation programmes need not be over-emphasized.

The chief agent of soil erosion is water. It is estimated that about 5333 million tonnes of soil (amounting to about 16 tonnes/ha) is eroded from Indian territory. Nearly 29% of this is carried into the sea and nearly 10% is deposited in surface reservoirs resulting in loss of storage capacity of reservoirs and occurrence of floods. Conservation of water virtually reduces the soil erosion. Thus the scope of conservation is very wide, encompassing all the measures of soil conservation, moisture retention and nutrient preservation in the soil.

5.4 SOME ANCIENT INDIAN METHODS OF WATER CONSERVATION

The Indus Valley Civilization, that flourished along the banks of the river Indus and other parts of western and northern India about 5,000 years ago, had one of the most sophisticated urban water supply and sewage systems in the world. The fact that the people were well acquainted with hygiene can be seen from the covered drains running beneath the streets of the ruins at both Mohenjodaro and Harappa. Another very good example is the well-planned city of Dholavira, on Khadir Bet, a low plateau in the Rann in Gujarat. One of the oldest water harvesting systems is found about 130 km from Pune along Naneghat in the Western Ghats. A large number of tanks were cut in the rocks to provide drinking water to tradesmen who used to travel along this ancient trade route. Each fort in the area had its own water harvesting and storage system in the form of rock-cut cisterns, ponds, tanks and wells that are still in use today. A large number of forts like Raigad had tanks that supplied water.

In ancient times, houses in parts of western Rajasthan were built so that each had a rooftop water harvesting system. Rainwater from these rooftops was directed into underground tanks. This system can be seen even today in all the forts, palaces and houses of the region.

Underground baked earthen pipes and tunnels to maintain the flow of water and to transport it to distant places, are still functional at Burhanpur in Madhya Pradesh, Golkunda and Bijapur in Karnataka, and Aurangabad in Maharashtra.

5.5 METHODS OF WATER CONSERVATION

There are many ways to conserve water. We need to always be thinking about conserving water. If we are thinking about it, we are less likely to be wasting the water that we use. On this web page I have listed several ways to conserve water indoors, outdoors, and also in landscaping. This is something that we must take seriously. It is our responsibility to conserve water whenever possible. We as Americans waste gallons of water every day. We must learn to save our water so that in the years ahead of us, we will have plenty of pure, clean water.

5.5.1 Technical Methods to Conserve Water

Rainwater Harvesting : Rainwater harvesting is the gathering and collection of water from the rooftop. The traditional method of rain water harvesting is the most effective and simple way to conserve the water. It means utilization of rain water for the domestic as well as agricultural purposes. There are three technical methods of rain water harvesting such as Catchment, Conveyance and storage.

Historical Water Bodies : There are many traditional water bodies which have been in disuse for the longer time. These bodies can be reused as the recharging points.

Ponds : Steps should be taken to avoid dumping of sewage into the village ponds. Efforts need to be made to deepen these ponds with the dragline machines. Garbage and other waste sholud not be dumped into the ponds.

5.5.2 In House Conservation of Water

Home water use varies considerably, depending upon the number of people in a household, plumbing fixtures, appliances, and other factors. The largest water users in the home are toilets, clothes washers, faucets, and showers.

Water supply planners estimate that a typical household needs 0.4 to 0.5 acre-feet of water per year (approximately 150,000 gal) to satisfy the typical demands of a home and lawn. However, we can get by on far less.

Outdoor water use accounts for about 55 percent of the residential water use in urban areas along the Front Range, most of which is used on turf. As a percentage of total water use in the urban Front Range, outdoor water accounts for about 40 percent of all urban water use. Many Colorado residents use over 200 gallons of water per capita per day. In contrast, residents in Arizona use 160 gallons per capita per day (20 percent less than most Colorado residents). Water conservation is vital to enhancing the efficiency of how we use water. This fact sheet describes ways to reduce water consumption in and around your home.

Dishwasher Efficiency

Use your automatic dishwasher only when you have a full load. Your dishwasher uses the same amount of water for a small load of dishes as it would for a full load.

- Install a high efficiency dishwasher machine.
- Running the dishwasher only when it's full can save 1,000 gallons of water per month.
- Running a full dishwasher usually uses less water than washing the same number of dishes by hand.
- Because the drying cycle of most dishwashing machines uses 1,500 watts per cycle, air or hand drying the dishes is more efficient and less expensive.
- Additional ways to conserve water in the home:
- Check your water meter and bill and talk to family members about setting water conservation goals for indoor water use.
- Use the garbage disposal less often. Compost organic matter from your kitchen.
- Collect the water you use for rinsing fruits and vegetables and reuse it to water houseplants and/or shrubs.
- When buying a new appliance, remember that certain models offer different cycles that are more water and energy-efficient.
- If you have an evaporative air conditioner, direct the water drain to a flower bed, tree, or your lawn.
- Collect rainwater in a bucket for watering indoor plants.
- If your local water utility offers incentives or in-home water audits, take advantage of these programs.
- Keep drinking water in the refrigerator during the summer instead of letting the faucet run until water is cool.

Kitchen Sink

Rather than using an electric disposal for kitchen waste, start a compost pile.

Kitchen Tips

1. Install a low-flow faucet aerator, which can cut water use in half.
2. Soak pots and pans before washing. When washing dishes by hand, fill one sink or basin with soapy water.

3. Fill the basin or a pan with water to wash fruits and vegetables.
4. Keep a pitcher of water in the refrigerator rather than running tap water until it is cool enough to drink.
5. When buying a new dishwasher, consider purchasing a water-saving model. Newer models can cut water use by 25 percent and are no more expensive than non-conserving models.
6. Wash only full loads in the dishwasher.

Washing Machine

When using your automatic washing machine, adjust the water level to the correct setting for the size of load you plan to wash.

Laundry Room Tips

1. When buying a new clothes washer, consider purchasing a water-saving model. New horizontal axis models can save up to 40 percent of the water used by a conventional model. Check with your municipality to see if they provide rebates on the purchase of water-saving clothes washers.
2. Wash only full loads in the clothes washer.
3. Insulate your water pipes. You'll get hot water faster plus avoid wasting water while it heats up.

Don't Leave Water Running

If you wash dishes by hand, don't let the water run down the drain. If you have two sinks, fill one with soapy water and one with water for rinsing. If you have only one sink, gather washed dishes in a dish rack and rinse them with a spray device or a pan full of hot water.

Drinking Water

Try keeping a pitcher or bottle of water in the refridgerator for cool drinking water. Running tap water to cool it off for drinking water is wasteful. Also, if you don't drink all of the water in your glass, rather than dumping the rest down the drain, use it to water a plant.

Check for Leaks

Often faucets and pipes can have leaks and we don't even notice. Even the smallest leak can waste up to 20 gallons of water a day. Leaks waste water 24 hours a day, seven days a week, and can often be repaired with only an inexpensive washer.

Check your Toilets for Leaks

Put a little food coloring into your toilet tank. If the color begins to appear in the bowl, without flushing, you have a leak that should be repaired as soon as possible. Install a low-flow or high-efficiency toilet to save nearly 2 gallons of water per flush.

Don't Use the Toilet as a Waste Basket

Every time you flush a facial tissue or other small bit of trash, you waste five to seven gallons of water.

Showers

Take shorter showers. Long, hot showers can waste five to ten gallons every unneeded minute. Limit your showers to the time it takes to soap up, wash down, and rinse off. You can also install a water-saving shower head to help reduce the amount of water used during a shower.

Don't run the shower for five minutes to warm it up while you are doing something else - adjust the temperature when you get in. And keep your shower time to five minutes. Turn the shower off if you need to shave or deep condition your hair.

In addition to shorter showers, install a low-flow showerhead, which will also reduce water waste. While you're at it, repair your leaky faucets.

Showerheads currently manufactured in the U.S. have a flow-rate of 2.5 gallons per minute (gpm) or less. Here are some suggestions for increasing shower-use efficiency.

- Install a low-flow showerhead if you do not already have one.
- Keep your showers brief. A shower that lasts for five minutes using a low-flow showerhead uses 12 gallons of water. If possible, use a watch to time yourself while you are in the shower.
- Turn off the water while you lather up with soap and shampoo.
- Irrigate your indoor plants by placing a bucket in the shower to collect the water while waiting for it to warm up.
- Check the flow rate of your showerhead by using a 5-gallon bucket and a clock. Turn the shower on full and place a 5-gallon bucket under the shower for two minutes. A 2.5 gpm showerhead will fill the bucket up in that two-minute time frame.
- Check the flow rate of your showerhead by using a 5-gallon bucket and a clock. Turn the shower on full and place a 5-gallon bucket under the shower for two minutes. A 2.5 gpm showerhead will fill the bucket up in that two-minute time frame.
- Check and repair leaks in the tub diverter valve.

Plant Sense

Change your landscaping and house plants to low-water, native or drought-tolerant plants. Then do your watering in the morning or evening when there is less chance of evaporation. Additionally, use your broom instead of the hose to clear debris from your driveway and side-walks.

Become a Flexitarian

The food production water usage for plant-based foods is significantly less than that of meat-based foods. A plant-based diet is healthier, too. (Click for more information on The Flexitarian Diet.)

Install a Water Filtration System

Instead of buying bottle after bottle of drinking water, install a water filter and use refillable beverage containers.

Install Energy-Efficient Appliances

Use an EnergyStar washing machine and dishwasher and only run full loads, or at the very least use the lowest water level setting you need for smaller loads.

Use a Bucket

Wash your car with a bucket of soapy water instead of letting the hose run.

Set up a Rain Water Barrel

Collect rain water and use it later to water the garden. Another option is to install a gray water system.

Use Biodegradable Cleaning Products

Take advantage of the growing selection of eco-friendly household cleaning products, dish washing soaps, and laundry detergents. Biodegradable products won't pollute the water.

Bathroom Water Efficiency

Take baths instead of showers. A bath in a partially filled tub uses less water than all but the shortest showers.

Toilets made before 1993 use 3.5 to 8 gallons per flush (gpf). High efficiency toilets manufactured after 1993 use 1.6 gpf or less. The date of manufacture of most toilets is on the underside of the tank lid. A family of four can save 14,000 to 25,000 gal/yr by switching from conventional toilets to the newer, more efficient ones. Your water utility may even offer rebates for replacing convential toilets. Additional water savings can occur by making sure your toilet is not leaking and that the flapper is working properly. Here are other suggestions for increasing your toilet-use efficiency.

- Install vacuum assisted, low-volume toilets.
- Consider not flushing the toilet unless absolutely necessary.
- Regularly check for toilet leaks by placing food coloring in your toilet tank. Repairing leaking toilets can save more than 600 gallons of water per month.
- Do not use your toilet as a wastebasket.
- Make sure your toilet flapper does not remain open after flushing.

- Avoid using toilet bowl cleaners such as toilet tank tablets. These products affect the pH of water in your toilet tank and can cause leaks by damaging the rubber and plastic parts of your toilet.

Bathroom Tips

1. Bathroom use accounts for about 65 percent of the water used inside the home.
2. Check regularly for any leaks and fix them. Most common bathroom leaks are found in faucets and in and around toilets.
3. Replace older, larger-use toilets with the newer ultra-low flush models. Standard toilets manufactured prior to the 1980s usually require 15 to 20 litres per flush. Toilets sold during the 80s and early 90s use 13 litres per flush.
4. Do NOT use the toilet to dispose of paper, facial tissues, or cigarettes.
5. Take a five-minute shower.
6. Use the minimum amount of water needed for a bath by closing the drain and the filling the tub only 1/3 full.
7. Install a low-flow showerhead. It can save about half the amount of water you typically use in the shower, while still providing a refreshing, cleansing shower.
8. Turn the tap water off while brushing your teeth, shaving, or washing your face.
9. If the toilet flush handle frequently sticks in the flush position, letting water run constantly, replace or adjust it.

Faucet Efficiency

- Install low-flow faucet aerators on all your household faucets. Some aerators can restrict flow to less than 1.0 gpm.
- Do not run the faucet continuously while washing dishes and hands, shaving, or brushing your teeth.
- Checking and repairing faucet leaks can save up to 140 gallons of water per week.

Estimated Facet Leakage Rates (# of drips)

60 drops/minute = 192 gallons/month

90 drops/minute = 210 gallons/month

120 drops/minute = 429 gallons/month

Clothes Washing Efficiency

Conventional washing machines use between 35 to 50 gallons per load (gpl). The newer front-loading machines are more efficient and use between 18 to 20 gpl. Below are suggestions for reducing water use while clothes washing.

- Run the washing machine only when you have a full load of clothes.
- For lightly soiled laundry loads, use the shortest wash cycle.
- To avoid redundant washing, pre-treat stains on your clothes.
- Select the minimum water volume per load if your washer has a variable water volume setting.
- Regularly check washing machine hoses for leaks.

Brushing Your Teeth

Turn off the water while brushing your teeth. There is no need to keep water pouring down the drain. Just wet your brush and turn off the water. You can turn the water back on when you are ready to rinse.

Shut the Water Off

Instead of letting the water run while you brush your teeth, shave, peel vegetables or scrub dishes, turn it off until you need it.

5.5.3 Outside Conservation of Water

Watering

Don't waste water by letting it fall on areas where it is not needed. Position your sprinklers so water lands on the lawn or garden, not on paved areas. Also avoid watering on windy days.

Do you Need to Water?

Water your lawn only when it needs it. A good way to see if your lawn needs watering is to step on the grass. If it springs back up when you move, it doesn't need water. If it stays flat, then it needs to be watered.

How to Water?

Deep soak your lawn. When you do water, do it long enough for the moisture to soak down to the roots where it will do the most good. A light sprinkling can evaporate quickly and tends to encourage shallow root systems.

When to Water?

Water during the cool parts of the day. Early morning generally is better than dusk since it helps prevent the growth of fungus.

Managing the Water Needs of Plants

Plant drought-resistant trees and plants. Many beautiful trees and plants thrive with far less watering than other species.

- Reset automatic controllers according to the seasonal needs of plants. Inspect controls at least once a month to adjust run times.
- Winter watering will minimize stress to trees, shrubs, flowers, and turf in areas receiving low winter precipitation. Apply water once a month during dry winter periods.
- Drip irrigation installed at the base of large transplanted trees will need to be moved as the tree root system expands. At maturity, tree roots spread outward two to three times the width of the tree canopy.
- Shaded plants use less water than plants in full sun.

Planting Tips

1. Add two to four inches of organic material, such as peat or compost, to the soil. Greater soil depth will increase the ability of the soil to retain moisture.
2. Use water-wise plants. Native and adaptive plants will use less water and be resistant to local plant diseases and pests.

Use Mulch

Put a layer of mulch around trees and other plants. Mulch will slow the water from evaporating, and discourage weed growth, too.

Cleaning Driveway and Sidewalk

When you use a hose to clean the driveway or sidewalk, you are washingsediment and poisons into storm drains which flow directly into rivers. Use a broom to clean driveways and sidewalks so that you can dispose of this waste properly.

Washing the Car

Don't let the water run while washing your car. Clean the car with a pail of soapy water. Use the hose just to rinse it off.

Using a running hose to wash your car can waste about 400 litres of water. Using a bucket with a sponge plus a trigger nozzle on the hose will save you about 300 of those litres.

Pool

If you own a pool, be sure to use a pool cover when it's not in use. This will cut down on evaporation losses and will keep it cleaner and warmer. Check equipment such as filtration systems and water inlets on a regular basis for signs of leaks.

Rain Barrel

Collect rain water in an old barrel or other large container that is outfitted with a spigot and a suitable cover, and use the water on your garden. Use this water as an alternative to turning on the hose for newly transplanted material, window boxes, flower pots and container gardens. Rain water is actually better for your plants as it does not contain any chlorine and is at ambient temperature. Keep your rain barrel covered to prevent mosquitos breeding and for safety reasons.

Hoses and Sprinklers

Don't play with hoses and sprinklers. These devices are for watering the lawn and garden. They are not toys and they use a lot of water.

Check for Leaks

Check for leaks in pipes, hoses, faucets, and couplings. Leaks outside the house may not seem as bad since they're not as visible, but they can be just as wasteful as leaks inside. Check frequently and keep them drip-free.

By Gardening

Many people in Washington state enjoy growing their own vegetable garden or flower garden. Growing your own garden is a wonderful experience, but there are many responsibilities that come with growing your own garden. These responsibilities need to be taken very seriously.

Many people who use pesticides and fertilizers on their lawns and gardens do not realize the kind of damage they are doing to our natural environment. We all need to realize that we can have a beautiful lawn or garden without using pesticides, herbicides, or fertilizers which harm people, wildlife, and the environment. Some of the ways that you can help include organic gardening,composting, and landscaping with native plants.

Organic gardening is the best way to avoid contaminating our water. Organic gardening means growing plants in soil that is uncontaminated by chemicals and contains organic, natural substances which provide a healthy, natural habitat for the plants. The ultimate purpose of organic gardening is to provide a source of food which is uncontaminated by foreign substances (chemicals - pesticides,herbicides, and fertilizers), and will replenish itself over time. This is achieved by paying attention to soil quality, beneficial insects, and barrier plants. It includes animals, insects, plants and soil which are all necessary parts to maintaining an ideal organic garden that will be replenishing.

You cannot grow a garden without water. All living things are dependent upon water. That is why it is so important that we all realize the potential of conserving water. When we all work together to conserve our state's water, we can make a big difference. If we allow our water to continue to be polluted by chemicals and other poisons, it could mean permanent damage to our food supply.

- Plant in blocks instead of rows to create shade for the root systems and reduce evaporation.
- Group plants with similar water needs together.

- Check the soil for moisture before you water and do not water until the soil has dried out to a depth of at least 4 inches.
- Control weeds that compete with vegetables for water.

Improving Lawn Irrigation Efficiency

Urban lawn watering is the single largest water demand on most municipal supplies. However, there are many ways to conserve water on the home landscape.

First, create an irrigation schedule relative to the types of plants in your landscape. Second, learn the water requirements of your landscape plants and water accordingly, avoiding over-watering as much as possible. Over-watering is not only wasteful, it is also unhealthy for plants. Third, make a habit of manually operating your irrigation system and rely less on the automatic controller. Fourth, do not irrigate on a set schedule, since daily plant water use can vary greatly according to the weather. Finally, take into account recent rainfall amounts before watering your landscape.

During drought or times of restricted landscape watering, most lawns, including bluegrass, will withstand reduced watering regimes by going dormant. In these situations, adjust mowing, fertilizing, aeration, and weed control practices appropriately to the watering schedule. Most lawns can be revived with good management and care after the drought breaks. Changing landscape plants and lawn grass species during drought is not a good idea, as it generally takes more water to establish new plants than to keep old plants alive.

Keep in mind the variety of water needs among various landscape plants. For example, certain types of lawns may need water every three or four days during a hot, dry summer. However, trees and shrubs may only need water every few weeks, while flowerbeds may need to be watered once a week. Trees, shrubs, and flowers may rot if you water them on the same watering schedule as your lawn.

Guidelines for when to water:

- Irrigate when footprints or mower tracks become visible and/or large areas of the lawn become blue-gray in color.
- Slowly apply irrigation water at rates that replace evaporation (ET) so runoff and puddling do not occur. Water less in cool spring and fall weather and more (3/4 to 1 inch) during hot summer months.
- Move your sprinkler around the yard in cycles to let the water thoroughly and evenly soak in.
- Water dry spots instead of the entire lawn.
- Water between 9 p.m. and 9 a.m. to avoid evaporation losses from hot and windy weather conditions.

5.5.4 Additional Ways to Conserve Water for Landscaping

- Make sure the irrigation system is operating properly.
- Replace broken or missing sprinkler heads.
- Make sure the spray heads turn properly.
- Adjust heads so that water does not reach streets and driveways.
- Check nozzles for plugging.
- Place straight-sided containers (such as tuna fish cans) around the yard during irrigation and measure water depth so that you know how long it takes to apply ¼ to ½ inch of water.
- Place containers on persistent dry spots to determine if poor sprinkler coverage is the problem.
- Never water if the soil is still wet.

Outdoor Tips

1. In the summer, lawn watering and other outdoor uses can account for up to 50 percent of home water use. Studies show that as much as half of this outdoor use is wasteful. As a general rule, 2 to 3 cm of water per week is adequate.
2. Don't overwater your landscape. It can cause yellowing leaves or poor plant health. Give plants only the amount of water that they need.
3. Use low-angle or pulsating sprinklers that produce large fat droplets of water. Sprinklers that spray the water high into the air or produce a mist or fine spray lose much of the water through evaporation.
4. Set sprinklers to water the lawn, not sidewalks and driveways.
5. Check your sprinkler or irrigation systems regularly for any leaks, and fix them.
6. Be sure your hose has an automatic shutoff nozzle to ensure water is not wasted when the hose is left unattended.

5.5.5 By Landscaping

Wherever you live, there are plants that are native to your environment. These plants are accustomed to the conditions where you live because they have adapted to be able to survive on their own. Lawn and landscape maintenance often requires large amounts of water, but using native plants could reduce the amount of water that you need. When you landscape with the plants that arenative to your area, you are helping to conserve water because native plants don't need as much water as some other plants that are not accustomed to the climate and other conditions of your area.

There are several ways that landscaping with native plants helps theenvironment. When you landscape with native plants, you will need less water. When you don't plant native plants, then you have to water more, which causes water pollution. Also landscaping with native plants provides a habitat for endangered native plants and the creatures that have always depended on them.

Landscaping with native plants reduces your need to use fertilizers andpesticides. Plants that are not native to our climate and precipitation will not survive without help from us. When we try to keep a plant alive and well that is not native to our area, we cause several problems for theenvironment. We have to use fertilizers, pesticides, and gallons of water just to keep them alive. By using native plants, when landscaping, you can reduce the amount of work needed to keep things looking nice, as well as reducing the amount of fertilizers, pesticides, and water you might need.

By landscaping with native plants, you are helping to preserve the naturalhabitat for the diminishing number of native plants in the world. Also, by landscaping with native plants, you are supporting the wildlife and nativeinsects that depend on them for survival. If all of the native insects and plants die out, it may make all of the non-native pests a bigger problem.

It is helpful to schedule a time early in the morning or in the evening for watering your lawn and garden. You should avoid watering in the heat of the afternoon, and you also shouldn't water more than about once a week. A good way to tell if your lawn needs watering is to step on it and when you move your foot if the grass stands back up you're good to go, but if it stays smashed down then you probably need to water. Using mulch can effectively help reduce the amount of water that your plants need. Mulch helps keep water down in the ground, where the plants need it, for a longer period of time. Mulch will also help reduce the amount of weeds that could choke out your plants. Also, weeds sometimes absorb all of the water that you might have intended for your plants so you should try to keep as many weeds out as you can.

5.5.6 Rainwater Harvesting

In urban areas, the construction of houses, footpaths and roads has left little exposed earth for water to soak in. In parts of the rural areas of India, floodwater quickly flows to the rivers, which then dry up soon after the rains stop. If this water can be held back, it can seep into the ground and recharge the groundwater supply.

This has become a very popular method of conserving water especially in the urban areas. Rainwater harvesting essentially means collecting rainwater on the roofs of building and storing it underground for later use. Not only does this recharging arrest groundwater depletion, it also raises the declining water table and can help augment water supply. Rainwater harvesting and artificial recharging are becoming very important issues. It is essential to stop the decline in groundwater levels, arrest sea-water ingress, i.e. prevent sea-water from moving landward, and conserve surface water run-off during the rainy season.

Town planners and civic authority in many cities in India are introducing bylaws making rainwater harvesting compulsory in all new structures. No water or sewage connection would be given if a new building did not have provisions for rainwater harvesting. Such rules should also be implemented in all the other cities to ensure a rise in the groundwater level.

Realizing the importance of recharging groundwater, the CGWB (Central Ground Water Board) is taking steps to encourage it through rainwater harvesting in the capital and elsewhere. A number of government buildings have been asked to go in for water harvesting in Delhi and other cities of India.

All you need for a water harvesting system is rain, and a place to collect it! Typically, rain is collected on rooftops and other surfaces, and the water is carried down to where it can be used immediately or stored. You can direct water run-off from this surface to plants, trees or lawns or even to the aquifer.

Some of the benefits of rainwater harvesting are as follows

- Increases water availability
- Checks the declining water table
- Is environmentally friendly
- Improves the quality of groundwater through the dilution of fluoride, nitrate, and salinity
- Prevents soil erosion and flooding especially in urban areas

5.5.7 Agriculture

Conservation of water in the agricultural sector is essential since water is necessary for the growth of plants and crops. A depleting water table and a rise in salinity due to overuse of chemical fertilizers and pesticides has made matters serious. Various methods of water harvesting and recharging have been and are being applied all over the world to tackle the problem. In areas where rainfall is low and water is scarce, the local people have used simple techniques that are suited to their region and reduce the demand for water.

- In India's arid and semi-arid areas, the 'tank' system is traditionally the backbone of agricultural production. Tanks are constructed either by bunding or by excavating the ground and collecting rainwater.
- Rajasthan, located in the Great Indian Desert, receives hardly any rainfall, but people have adapted to the harsh conditions by collecting whatever rain falls. Large bunds to create reservoirs known as khadin, dams called johads, tanks, and other methods were applied to check water flow and accumulate run-off. At the end of the monsoon season, water from these structures was used to cultivate crops. Similar systems were developed in other parts of the country. These are known by various local names-- jal talais in Uttar Pradesh, the haveli system in Madhya Pradesh, ahar in Bihar, and so on.

5.5.8 Reducing Water Demand

Simple techniques can be used to reduce the demand for water. The underlying principle is that only part of the rainfall or irrigation water is taken up by plants, the rest percolates into the deep groundwater, or is lost by evaporation from the surface. Therefore, by improving the efficiency of water use, and by reducing its loss due to evaporation, we can reduce water demand.

There are numerous methods to reduce such losses and to improve soil moisture. Some of them are listed below.

- Mulching, i.e., the application of organic or inorganic material such as plant debris, compost, etc., slows down the surface run-off, improves the soil moisture, reduces evaporation losses and improves soil fertility.
- Soil covered by crops, slows down run-off and minimizes evaporation losses. Hence, fields should not be left bare for long periods of time.
- Ploughing helps to move the soil around. As a consequence it retains more water thereby reducing evaporation.
- Shelter belts of trees and bushes along the edge of agricultural fields slow down the wind speed and reduce evaporation and erosion.
- Planting of trees, grass, and bushes breaks the force of rain and helps rainwater penetrate the soil.
- Fog and dew contain substantial amounts of water that can be used directly by adapted plant species. Artificial surfaces such as netting-surfaced traps or polyethylene sheets can be exposed to fog and dew. The resulting water can be used for crops.
- Contour farming is adopted in hilly areas and in lowland areas for paddy fields. Farmers recognize the efficiency of contour-based systems for conserving soil and water.
- Salt-resistant varieties of crops have also been developed recently. Because these grow in saline areas, overall agricultural productivity is increased without making additional demands on freshwater sources. Thus, this is a good water conservation strategy.
- Transfer of water from surplus areas to deficit areas by inter-linking water systems through canals, etc.
- Desalination technologies such as distillation, electro-dialysis and reverse osmosis are available.
- Use of efficient watering systems such as drip irrigation and sprinklers will reduce the water consumption by plants.

5.5.9 Other Resources

There are many other resources for learning about water conservation. Please find below links to other web sites, titles and authors of books, and also some titles of some really informative videos. There is so much that can be learned about water conservation and its importance. It is very important that we all take our responsibility and conserve the water so that future generations will have plenty of clean, pure water.

The Adoption of Water Conservation Application Techniques

A number of farm application irrigation technologies that could contribute to the conservation of water in both surface and piped systems have been developed. These techniques are generally adopted for the cultivation of high value crops. About 20 million hectares worldwide are now irrigated by sprinkler irrigation. Micro-irrigation as a technology has matured into reliable water and fertigation management system for crop production over the past decades and its usage continues to increase rapidly. A total of 3.2 million hectares are being irrigated by micro-irrigation techniques throughout the world, which represent one percent of the total areas irrigated in the world. A growth of more than six times over the last 20 years has been achieved, which is quite a success. Micro-irrigation is an important tool in the drive towards food security. Micro-irrigation systems have great potential to create significant opportunities for smallholder agriculture.

During the 1980s micro-irrigation started penetrating developing countries, mainly India. A few years later, fearing a water and food shortage, the Chinese Government declared drip irrigation technology to be amongst the top priorities for development in the last decade. However, Vidal et al., (2001) have shown that micro-irrigation "is not a miracle technology: excellent as well as poor results were obtained from these technologies, and their adoption really depends on farmers capacity to finance and operate them, as well as on the type of crop production".

According to the ICID database on modern irrigation techniques, it appears that the use of pressurized water application techniques is widely different between countries. The percentage of modern technique decreases with the total area irrigated in developed countries from a low of 21 percent in the USA and 16 percent in Italy to 90 percent or over in Northern European countries (France and Germany). Sprinkler has a much greater application compared to that of the drip systems with the exception of Cyprus, Israel and Jordan.

In South Asia and Africa low cost bucket-and-drip sets are becoming increasingly popular with farmers. In areas where shallow groundwater is plentiful, thousands of poor farmers in Bangladesh have used low-cost treadle pumps to supply water for crops for their own food security and additional income.

Advanced surface irrigation techniques such as precise land levelling, gated pipes or cablegation have also been developed to improve the typical poorly distributed application of water under conventional methods by furrow, border or basin irrigation. Experience shows that these techniques are too sophisticated to be adopted by most smallholders in developing countries. Adaptive research in the Middle East and North Africa has shown the difficulties in introducing surface water conservation techniques. Many attempts to introduce gated pipes in Morocco for example have failed for a number of reasons, mainly the poor maintenance of the quaternary earth canals. Considerable efforts for training of farmers in water conservation in surface systems are needed.

Farmers have adopted pressurised systems much faster than surface application techniques. For example, about 65 percent of the farmers in the Jordan valley have shifted from surface to drip irrigation over a 10-year period. Farmers built on-farm storage reservoirs to provide the flexibility required for drip irrigation.

Although there are no statistical data available on it, it is likely that most of the areas irrigated through water conservation techniques are served from wells or piped systems. There is still a huge potential for development of water conservation techniques in the areas presently irrigated from groundwater. Application of these techniques will continue to expand with the market for high value crops. However, the adoption of these techniques is limited to the cultivation of certain crops because of technical and financial considerations.

5.6 WATER SITUATION IN INDIA: AN OVERVIEW

India as a nation uses 500 Billion Cubic Meters (BCM) of water every year for various purposes, most important of which are irrigation and domestic use. This equates with a per capita use of around 470 cu meters per year, which is far short of the amount used in a developed country like United Stated where per capita water use is around 1600 cu meters. But then, United States has a larger territory, and a far smaller population. The lower per capita use also denotes a scarcity that is likely to become much worse by year 2025 due to population growth and increasing urbanisation.

India receives a rainfall of around 4000 BCM every year, actually enough to take care of India's water needs, but most of this water is received within 4 months of 'monsoon' season between June and September, or be more precise, during less than 100 hours of precipitation spread over these four months. Most of this water is lost. A small part of it goes into soil and rechargesground water, but most of it evaporates and is lost.

Traditionally, till the middle of twentieth century, India largely managed its water needs by its major rivers, and rainfall, supplemented by traditional water bodies called by different names like 'johad', 'talab', bavdi, or 'jheel' - most of which were nothing but natural or man-made ponds that collected water during rainfall that was used by surrounding villagers for various purposes. After the great demographic divide of 1919, after which the population of India started exploding, these sources were unable to satisfy the needs. As a result more and more people resorted to drilling for groundwater.

Today, the groundwater levels are falling everywhere throughout the country. In some cities they are falling at an alarming rate of almost 10 meters every year, indicating a crisis. The realisation that ground water will not last forever has resulted in a strong demand for finding solutions to water scarcity that threatens the country. Though there is a realisation for water conservation policy, a comprehensive policy in this regard is yet to emerge.

5.7 WATER CONSERVATION POLICY AS A SOLUTION TO WATER SCARCITY

With a rainfall that is adequate to take care of its needs, there is a realisation that water conservation is the only answer to India's problems. Some cities have made rules making it compulsory to have water harvesting systems in all new buildings. Many Non-Governmental Organisations (NGOs) have taken up projects related to water conservation in rural and semi-arid areas with excellent results. One such NGO, 'Tarun Bharat Sangh' has been involved with people and by way of self-help, it has created 3000 water conservation structures in 650 villages. However, a lot more needs to be done in this regard.

Different experts and stakeholders have suggested different methods, but there are some that everyone agrees as a solution.

(i) **Local Water Bodies** - This used to be one of the main local sources of water earlier, but gradual filling of those bodies and their pollution by people and industries have made most of them dysfunctional. Now their renovation is looked upon as one of the means to store and conserve rain water.

(ii) **Urban Water Harvesting** - As India gradually shifts from rural to urban areas, the role of water conservation in urban areas is increasingly recognised and accepted as a solution towater problem.

(iii) **Afforestation** - Planting of trees has been accepted as a single biggest means of holding water in a territory, and is increasingly resorted to as a means especially in the semi-arid regions. However most of the efforts are not well coordinated till now in this regard.

(iv) **Improving Technology & Water Use Methods** - Most Indian farmers are still not fully acquainted with efficient water usage methods for irrigation. Methods like drip irrigation are still unknown in most parts even where water is scarce. Awareness in this regard needs to grow to allow efficient use of water.

(v) **Water Pricing As Means To Minimize Wastage** - So far, the governments have generally desisted from pricing water at its real cost. The unrealistically low prices tend to encourage overuse and wastage. The World Bank has been advocating pricing as a means to avoid wastage, but the democratic governments have so far not been inclined to accept this advice.

5.8 NEED TO PLAN FOR THE IMPENDING CRISIS

India stands today at a point where it faces a real danger of a water crisis if it is unable to plan for the solutions and start implementing them very soon. The most feasible solution is toconserve water by all the different means possible, and a comprehensive water conservation policy with necessary legal sanction for enforcement, and enough propaganda to create public awareness is an acute need of the hour.

Water conservation refers to reducing the usage of water and recycling of waste water for different purposes such as cleaning, manufacturing, and agricultural irrigation. Water conservation can be defined as:

1. Any beneficial reduction in water loss, use or waste as well as the preservation of water quality.
2. A reduction in water use accomplished by implementation of water conservation or water efficiency measures; or,
3. Improved water management practices that reduce or enhance the beneficial use of water.

5.8.1 Goals

The goals of water conservation efforts include as follows:

- **Sustainability:** To ensure availability for future generations, the withdrawal of fresh water from an ecosystem should not exceed its natural replacement rate.
- **Energy conservation:** Water pumping, delivery, and wastewater treatment facilities consume a significant amount of energy. In some regions of the world over 15% of total electricity consumption is devoted to water management.
- **Habitat conservation:** Minimizing human water use helps to preserve fresh water habitats for local wildlife and migrating waterfowl, as well as reducing the need to build new dams and other water diversion infrastructures.
- Reduce water consumption per capita.

5.9 IMPORTANT POINTS TO FOLLOW FOR WATER CONSERVATION

The most important step in the direction of finding solutions to problems of water and environment conservation is to change people's attitudes and habits which includes each one of us. Conserve water because it is the right thing to do. We can follow some of the simple things that have been listed below and contributed to water conservation.

I. Try to do one thing each day that will result in saving water. Don't worry if the savings are minimal because every drop counts ! You can make a difference.

II. Remember to use only the amount you actually need.

III. Form a group of water -conscious people and encouarage your friends and neighbours to be part of this group. Promote water conservation in community news letters and on bulletin boars. Encourage your friends, neighbours and co-workers to also contribute.

IV. Encourage your family to keep looking for new ways to conserve water in and around your home.

V. Make sure that your home is leak free. Many homes have leaking pipes that goods unnoticed.

VI. Do not leave the tap running while you are brushing your teeth or soaping your face.

VII. See that there are no leaks in the toilet tank. You can check this by adding colour to the tank. If there is a leak, color will appear in the toilet bowl within 30 minutes. (Flush as soon as the test is done, since food colouring may stain the tank.)

IX. Avoid flushing the toilet unnecessarily. Put a brick or any other device that occupies space to cut down on the amount of water needed in each flush.

X. When washing the car, use water from bucket and not hosepice.

XI. Do not throw away water that has been used for washing vegetable, rice or dals. Use it to water plants or to clean the floors, etc.

XII. You can store water in a variety of ways. A simple method is to collection source. You can also collect water in a bucket during the rainy season.

5.10 SOCIAL SOLUTIONS

Water conservation programs are typically initiated at the local level, by either municipalwater utilities or regional governments. Common strategies include public outreachcampaigns, tiered water rates (charging progressively higher prices as water use increases), or restrictions on outdoor water use such as lawn watering and car washing. Cities in dry climates often require or encourage the installation of xeriscaping or natural landscaping in new homes to reduce outdoor water usage.

One fundamental conservation goal is universal metering. The prevalence of residential water metering varies significantly worldwide. Recent studies have estimated that water supplies are metered in less than 30% of UK households, and about 61% of urban Canadian homes (as of 2001). Although individual water meters have often been considered impractical in homes with private wells or in multifamily buildings, the U.S. Environmental Protection Agency estimates that metering alone can reduce consumption by 20 to 40 percent. In addition to raising consumer awareness of their water use, metering is also an important way to identify and localize water leakage. Water metering would benefit society in the long run it is proven that water metering increases the efficiency of the entire water system, as well as help unnecessary expenses for individuals for years to come. One would be unable to waste water unless they are willing to pay the extra charges, this way the water department would be able to monitor water usage by public, domestic and manufacturing services.

Some researchers have suggested that water conservation efforts should be primarily directed at farmers, in light of the fact that crop irrigation accounts for 70% of the world's fresh water use. The agricultural sector of most countries is important both economically and politically, and water subsidies are common. Conservation advocates have urged removal of all subsidies to force farmers to grow more water-efficient crops and adopt less wasteful irrigation techniques.

New technology poses a few new options for consumers, features such and full flush and half flush when using a toilet are trying to make a difference in water consumption and waste. Also available in our modern world is shower heads that help reduce wasting water, old shower heads are said to use 5-10 gallons per minute. All new fixtures available are said to use 2.5 gallons per minute and offer equal water coverage.

5.10.1 Household Applications

Water-saving technology for the home includes:

- Low-flow shower heads sometimes called energy-efficient shower heads as they also use less energy,
- Low-flush toilets and composting toilets. These have a dramatic impact in the developed world, as conventional Western toilets use large volumes of water.
- Dual flush toilets created by Caroma includes two buttons or handles to flush different levels of water. Dual flush toilets use up to 67% less water than conventional toilets.
- Saline water (sea water) or rain water can be used for flushing toilets.
- Faucet aerators, which break water flow into fine droplets to maintain "wetting effectiveness" while using less water. An additional benefit is that they reduce splashing while washing hands and dishes.
- Wastewater reuse or recycling systems, allowing:
- Reuse of graywater for flushing toilets or watering gardens
- Recycling of wastewater through purification at a water treatment plant. See also Wastewater - Reuse
- Rainwater harvesting

- High-efficiency clothes washers
- Weather-based irrigation controllers
- Garden hose nozzles that shut off water when it is not being used, instead of letting a hose run.
- using low flow taps in wash basins
- Swimming pool covers that reduce evaporation and can warm pool water to reduce water, energy and chemical costs.
- Automatic faucet is a water conservation faucet that eliminates water waste at the faucet. It automates the use of faucets without the use of hands.

  Water can also be conserved by landscaping with native plants and by changing behavior, such as shortening showers and not running the faucet while brushing teeth.
- Use waste water for growth of plants

5.10.2 Commercial Applications

Many water-saving devices (such as low-flush toilets) that are useful in homes can also be useful for business water saving. Other water-saving technology for businesses includes:

- Waterless urinals
- Waterless car washes
- Infrared or foot-operated taps, which can save water by using short bursts of water for rinsing in a kitchen or bathroom
- Pressurized waterbrooms, which can be used instead of a hose to clean sidewalks
- X-ray film processor re-circulation systems
- Cooling tower conductivity controllers
- Water-saving steam sterilizers, for use in hospitals and health care facilities.
- Rain water harvesting.

5.10.3 Agricultural Applications

For crop irrigation, optimal water efficiency means minimizing losses due to evaporation,runoff or subsurface drainage while maximizing production. An evaporation pan in combination with specific crop correction factors can be used to determine how much water is needed to satisfy plant requirements. Flood irrigation, the oldest and most common type, is often very uneven in distribution, as parts of a field may receive excess water in order to deliver sufficient quantities to other parts. Overhead irrigation, using center-pivot or lateral-moving sprinklers, has the potential for a much more equal and controlled distribution pattern. Drip irrigation is the most expensive and least-used type, but offers the ability to deliver water to plant roots with minimal losses. However, drip irrigation is increasingly affordable, especially for the home gardener and in light of rising water rates. There are also cheap effective methods similar to drip irrigation such as the use of soaking hoses that can even be submerged in the growing medium to eliminate evaporation.

As changing irrigation systems can be a costly undertaking, conservation efforts often concentrate on maximizing the efficiency of the existing system. This may include chiseling compacted soils, creating furrow dikes to prevent runoff, and using soil moisture and rainfall sensors to optimize irrigation schedules. Usually large gains in efficiency are possible through measurement and more effective management of the existing irrigation system.

5.11 MINIMUM WATER NETWORK TARGET AND DESIGN

The cost effective minimum water network is a holistic framework/guide for water conservation that helps in determining the minimum amount of freshwater and wastewater target for an industrial or urban system based on the water management hierarchy i.e. it considers all conceivable methods to save water. The technique ensure that the designer desired payback period is satisfied using Systematic Hierarchical Approach for Resilient Process Screening (SHARPS) technique.

Few suggestions for improving conservation practices at the community level. These include:

- **Encouraging participation of beneficiaries:** Placing more emphasis at the community level and encouraging participation of people, particularly women and making the community as a whole realise the importance of efficient use of water.
- **Saving irrigation water:** The agricultural sector is the largest consumer of water and, therefore, needs to be focused for water conservation practices
- **Saving industrial water:** The industries should be encouraged to purify their water and to reutilise the once used water and to dispose off effluent safely, avoiding natural water resources.
- **Limiting evaporation:** Evaporation from open water bodies and reservoirs constitutes a major constraint in the efficient use of water. Suitable evaporation retardation or suppression practices must be adopted for limiting evaporation
- **Augmenting aquifer recharge:** Efforts at large scale artificial recharge can have the potential for storing water underground with minimum evaporation and surface storage facility.
- **Use of non-conventional water sources:** For example, treated municipal wastewater, agriculture drainage and brackish/ saline ground water etc., can be encouraged for supplemental use in agricultural irrigation after treatment.

- **Capacity building:** To develop skilled manpower, encourage field level training and live demonstrations of rational water use through modern irrigation methods, advice on soil moisture & crop water requirement etc.

Tips to save water :

- Avoid leakage of water from the taps.
- Turn the tap off when not in use especially when you brush your teeth or wash clothes.
- Rainwater harvesting is the another method to conserve water.
- The water supply should be limited in those areas which enjoys the unlimited water supplies.
- Check the leakage of water in the toilets. Also get check the hidden water leaks.
- Educate the mind of the people in the rural areas to save the water.
- Promote the conservation of water through media and wall posters.
- Never throw the water unnecessary on roads which can be used for gardening and cleaning.
- Avoid unnecessary flushing the toilets. Dispose off the tissues, cigarettes and other waste into the bin instead of toilets.
- Use minimum amount of water to bath.
- Water Waste restrictions.
- Improvement in the water distribution system.
- Water your lawn only when it is needed.
- Use a broom instead of hose to clean the sidewalks or to wash the car.
- Capture the water that is leaking and repair it as soon as possible.
- You can use washing machine to wash clothes that does not consume much water.
- Donot leave the tap running while washing the dishes in the kitchen.
- Install small shower heads to reduce the flow of water.

Benefits to Conserve water :

- If you save water it can save your money bills.
- Reduction in interior water use cuts waste water flows, especially overflowing of gutters which contaminates the environment.
- Environment benefits include eco system and habitat protection.
- Water conservation helps in improving the quality of your drinking water.

5.12 IMPORTANCE OF WATER TO PLANTS

The importance of water to plants, cannot be underestimated. In fact, without water, the plants cannot survive as their physiological activities will come to a standstill. Some examples are cited here.

1. The plants take their food nutrients in solution form. In the absence of water in the soil, nutrients become unavailable to plants.
2. Water in the plant cells help in the proper functioning of the cells.
3. With the movement of water inside the plant, food materials are also carried to different parts of the cells.
4. Water is used as a raw material in photosynthesis wherein the plant manufactures carbohydrate in the presence of light using water and carbon dioxide as raw materials.
5. Water influences the chemical, physical and biological activities of soil.
6. Plants transpire water and keep the temperature normal. Thus water controls the temperature of plant.
7. In winter, it prevents quick freezing of plants and protects them from frost.
8. Hormone and enzymes move inside the plant by dissolving in water.
9. Water is an integral constituent of the cell protoplasm. It keeps the cell turgid which is necessary for all physiological processes to operate.

5.13 IMPORTANCE OF WATER CONSERVATION

Water is very important content. Without water there is no life in earth. We use water in so many ways in our daily life. Water is a substance which covers ¾ part of the world. It means water is occupying more portion compared to land. But this water is becoming more polluted because of the environmental changes. So it is necessary to conserve the water. It is every one duty to protect the water because everyone uses this water and we the people are making water polluted. Now we may get the doubt how to conserve the water. A few simple changes in home about the water consumption make a vast contribution to water conservation.

The basic changes we have to do is stop leaking of water. We have to see whether there is a leakage in our taps. Because of the water leakage taps we can't protect the water. We can also protect the water by installing Low-Flow Showerheads. With these we can protect the water because it flows lower. And we won't use much water with these Low-Flow Showerheads. When we are brushing our teeth or when we are shaving we have to turn off faucets. After completing the work we have to use the water. By this change also we can preserve the water. Using sprinklers for yard is also a good remedy for water conservation. Turn off the water in the shower when shampooing the hair Water will be saved. It is better to replace the new toilets in place of old toilets. These are high efficiency and will use less water. To collect a rain water use a rain barrel at the bottom of gutter down spouts. It may use for watering plants. These are some of the things to follow to conserve the water.

We all use the water we must also see how we are using this. It is important to note that there are many reasons why we waste this natural resource; it has become necessary to waste the resources for innovating new technologies. However we have now proved it is not necessary to waste. It is easy now to conserve the water. By conserving water we can also save the money. There are hundreds of ways to conserve the water. But may not follow all those ways. We can follow only those ways which are easier to us. The easier things we can follow are the things which are mentioned above. They are the things we follow easily in our home.

It is the responsibility of every one to educate our children. With this we are indirectly conserving the water in future also. The children who are educating to save the water will save the water in future also. Now if we cut the usage of water it definitely helps the future generations. Water is mainly related to environment. Green also helps for the environment. Growing greenery everywhere and encouraging forest causes rain fall. So green is also conserving the water

5.14 NEED OF WATER CONSERVATION

Water is absolutely essential for utilization of all other natural resources including land. Agriculture depends on the availability and supply of water either from rainfall or from surface or underground fresh water resources. Most of the water is received through monsoon rains and is obtained from July to September. Only in some rivers of North India, water is received through the melting of snow in summer.

Because of high intensity of monsoon rains coupled with poor infiltration rates of soils, there is a heavy run-off of water. Run-off occurs when amount of precipitation is more than infiltration of water into the soil. Run-off is very heavy particularly in first rains. As a consequence of heavy run-off, the discharge in rivers increases during rainy period. The experience of last 2-3 decades, shows that every passing year the amount of run-off water discharged into the rivers is in creasing. It indicates that the amount of water which was conserved earlier in the site of rainfall (catchment area) has been reduced.

Our country receives more annual rainfall than the global average of annual rainfall. Therefore, water conservation has to be adopted as a national objective, not only for the sake of water availability but also for protecting the soil which is a precious resource.

5.15 WATER EROSION PROBLEM IN INDIA

Out of the total 328 m ha of geographical area of the country, about 143 m ha is cultivated and about 65 m ha is barren uncultivable waste and cultural wastelands. By and large these lands are more or less devoid of protective vegetation and physical barriers and hence are subjected to very serious water erosion. Nature takes 500 to 1000 years to develop 2.5 cm top soil but gets eroded in a course of a year or two. As a result it becomes the most harmful single factor in the deterioration of productive land. Along with the loss of run-off water a colossal amount of soil is also lost.

Overgrazing, deforestation, faulty cultivation, shifting cultivation and carelessly built roads in catchment areas have led to devastating down streams. These include gullying and floods leading to destruction of farm lands and villages, drop in flow during the dry season and consequent loss of crops, and siltation of reservoirs and canals. The problem has been further aggravated due to high rate of population growth: both human and livestock, resulting in indiscriminate exploitation of natural sources, for meeting the ever-increasing demand for fodder, fuel, fibre and fertilizers. Thus continuous degradation of production base and imbalance in land - water -plant-man-animal system is leading to ecological imbalance and economic insecurity through severe soil erosion both by water and wind and threat to the quality of our life and cultivation.

Even though no systematic survey for accelerating the extent of water and soil loss in different areas of the country, has been carried out, it. is estimated that about 5334 mt of soil is detached annually from its original place. The country's rivers carry 2052 mt of this, nearly 1572 mt is carried away by the rivers into the sea every year and 480 mt is being deposited in various reservoirs.

Soil erosion by water is mainly of four types:

1. Sheet erosion

   About 4-10 tonnes/ha/year in red soil, 17-43 tonne/ha/year in black soil and 4-14 tonne/ha/year in alluvial soil.

2. Gully erosion

   About 33 tonnes/ha/year in ravine regions

3. Hill side erosion

   More than 80 tonnes/ha/year in landslide, mine spoil areas etc.

4. Stream bank erosion

Along with soil, about 2.5 mt of nitrogen, 1.5 mt of phosphorus, 2.2 mt of potash, large amount of organic matter and micronutrients are also lost. The loss of nutrients is valued at about Rs.15000 million per annum. Mohanjodaro and Harappa are the examples of this fact. Glaring evidence of neglect of rainwater can be seen in the denuded hill tops all around and the extensive network of raville in the states of Gujarat, Madhya Pradesh, Uttar Pradesh and Rajasthan. Once a thick forest, Rajasthan is now a desert, Conservation of water is prerequisite to conserve soil which is a natural reservoir of plant nutrients. Taking into consideration the magnitude of problems, there is a great scope to streamline the entire water conservation programme more scientifically on micro watershed basis in cluster instead of sporadic efforts. Watershed management programme to conserve water and soil is a new approach and it is getting popularity very rapidly. The programme involves an appropriate combination of agronomic, mechanical, agrostological and afforestation measures, fitting to soil-climatic and topographical conditions of area and socio-economic conditions of the target group.

5.16 TYPES OF WATER LOSSES

Water losses with special reference to soil erosion aloe de- scribed below:

5.16.1 Raindrop or Splash Loss

Raindrops fall on the ground and form a splash along with suspended soil patlicles. Water and soil move away from their original place.

5.16.2 Sheet Erosion

When the intensity of rainfall is bit higher, a fairly uniform layer of soil is lost from the soil along with the run-off water. This type of soil and water loss is serious on slightly sloppy lands. The existence of this kind of soil and water erosion is seldom known to farmers as it does not leave behind any visible change on soil topography. Though the field appears the same as before but the cream of the soil i.e. highly fertile upper most layer, is skimmed off along with moving water. The estimates reveal that sheet loss of water erode away many tonnes of soil and has damaged millions of hectares of land in our country.

5.16.3 Rill Erosion

The speed of run-off water is sometimes high enough to create rills on the soil surface. Rills are very shallow finger sized channels which can be smoothened out by ploughing. Rill erosion is the intensified form of sheet erosion and more distinctly noticeable. The width of rills keeps on widening with every passing rainy season.

5.16.4 Gully Erosion

Gully erosion of water and soil is the aggravated form of rill loss. When rills are not smoothened out, water keeps on drawing through them and these shallow rills go deeper and deeper with every passing rainy season. The depth and size of rills go so high that they cannot be smoothened out with ploughing, they are called gullies. The size of gullies may be many metres wide and deep. They provide a big route to run-off water and a huge amount of water is lost through them.

5.17 FACTORS AFFECTING WATER LOSSES

Water in motion removes the soil and causes erosion of soil. Water erosion is due to dispersive action and transporting power of water. So it descends in the form of rain and leaves the land in the form of run-off. The intensity and speed of run-off are directly correlated to the amount of soil' and water loss. Some of the factors discussed below greatly influence the extent of soil and water loss.

5.17.1 Climate

The major climatic factors affecting run-off and erosion atе rainfall, temperature and wind. Of these, rainfall is the most important and the knowledge of rainfall characteristics is important in erosion studies. Raindrop impact on a soil surface has a pronounced effect on erosion. Raindrops cause soil splash, detach soil particles and make them available for transport. The amount of soil detached by the rain depends on the intensity of rain, its duration and the protective value of any cover present.

Erosive capacity of rain drops results from three factors, namely (1) amount and intensity of rainfall, (2) the diameter of raindrops, and (3) the velocity of drops as they strike the soil. The amount of soil transported by raindrops is governed by the transporting capacity of the raindrop energy; the case with which detached pat1icles are called; and the resistance afforded by the vegetation and other obstructions.

The amount, intensity, frequency and duration of rainfall have very definite effects on the amount and rate of resultant run-off. As large total. rainfall may not cause excessive erosion if the intensity is low. Likewise, an intensive rain of short duration may not produce sufficient run-off to cause erosion. When both amount and intensity factors are high in a given storm, run-off erosion will be serious.

5.17.2 Topography

Sloppy soils frequently permit run-off, resulting in the loss of water and soil, while on flat lands erosion is comparatively negligible. The degree of slope and length of slope are the main features of topography affecting erosion. Velocity of run-off water is influenced mainly by the degree of slope. If the land slope is increased four times, the velocity of water flowing over it is approximately doubled.

When the velocity is doubled the erosion capacity as represented by the kinetic energy of flowing water is increased about four times, tile amount of material of a given size that can be carried is increased by about 32 times, and the size of particles that can be transported by pushing or rolling is increased by about 64 times. There is a definite limit to the amount of suspended material that can be carried by given stream of water. When this limit is reached, more material cannot be picked up without increasing either the velocity or the depth. Any reduction in velocity on the other hand, will result in deposition of the suspended soil.

5.17.3 Vegetation

Vegetation is one of tile most efficient means to conserve water. A thick growth of grass or a dense forest may negate the effect of all those factors which accelerate water erosion. Vegetation reduces water erosion through the ways mentioned below.

Interception of Rainfall

The foliage of vegetation intercepts tile raindrops, retains a portion of water and evaporates it back into the atmosphere. However, the plant absorbs some of this water through foliage and utilizes it for metabolic purposes. The amount directly evaporated does not contribute to the run-off water.

Decreasing Run-Off Velocity

Vegetative cover serves as a hindrance to run-off water. A well distributed, close-growing vegetation not only slows down the rate at which water travels down the slope, but also tends to prevent " the concentration of water in a particular area. These two effects greatly reduce the erosive capacity of the run-off water.

Root Effect

Roots of grasses and other vegetation make the soil porous and permeable to water. The roots die and decay in the soil leaving behind numerous small cavities which increase the pore space and consequently the absorption of water.

Biological Effect

Remnants of vegetation serve as the food material for numerous organisms inhabiting the soil. These organisms range from large sized earthworms to microscopic bacteria, fungi, algae and actionmycetes. Macroorganisms work the soil and make it porous and permeable. Microorganisms decompose organic matter of the soil, and humus is formed as a decomposition product. Humus is very efficient in retaining water. It can retain water as much as 100 times of its own weight.

Soil

The amount of water left on the soil surface for run-off very much depends on soil properties. If the infiltration capacity of the soil is high, a large amount of water is soaked in by the soil and there is little run-off. In case soil is very compact or hardpan lies in the subsoil, a large amount of water is lost by run-off. The infiltration capacity or permeability of soil depends on number of factors such as soil structure, texture, organic matter content, moisture content and slope percentage.

5.18 AVENUES OF WATER LOSS FROM SOIL

Factors which influence behavior of water in soil in the context of soil erosion have been discussed above. Here, it is useful to briefly highlight the avenues through which water disappears from the soil.

5.18.1 Infiltration

The downward entry or movement of water into the soil is called infiltration. When the water is introduced to the soil either by irrigation or rain, it penetrates the surface of the soil. The speed of movement or the depth covered in a unit time is termed as infiltration rate. It is usually measured in centimeters per hour. The infiltration rate is greatly influenced by the moisture content of the soil. The soil takes in or absorbs water rapidly, when water is just applied on the field surface. If water supply continues, surface becomes saturated and infiltration rate decreases until it becomes merely constant.

The soil layer with the lowest transmission rate, either at the surface or below it, usually determines the infiltration rate. Surface sealing which is the formation of a thin compact layer on the ground surface rapidly reduces the rate of infiltration. The surface seal results from a breakdown in soil structure due to the beating action of raindrops and the action of the water flowing over the surface whereby, the fine particles are filtered around the larger particles to form a relatively impervious seal. Very sandy soils may have infiltration rates of over 25 cm per hour, whereas clay soils may have rates approaching zero. Poor soil structure, salinity and alkalinity reduce infiltration rate.

5.18.2 Run-Off

Run-off may be defined as the portion of rain or irrigation water which leaves the field either as surface flow or subsurface flow. As long as the rate of rainfall (or irrigation water) reaching the soil surface is less than the infiltration capacity, all the water is absorbed into the soil. When the rate of rain- fall exceeds the rate of infiltration, shallow depressions are filled to overflow level, water begins to flow over the land towards streams. The water required to fill depressions prior to the beginning of surface run-off is called initial detention or 'depression storage'.

Overland run-off usually reaches a tiny rivulet or channel which connects to the big rivers and ultimately to ocean. The time of travel from raindrop to stream channels depends on distance, slope, surface condition and depth of flow. Surface run-off is the run-off which travels over the ground surface to the channel of a watershed. The channels may range from rills to major streams. Subsurface flow is that portion of infiltrated water which penetrates to a shallow depth, travels laterally and is intercepted by channels or emerges to the surface. Appreciable amount of subsurface flow, relative to total run-off, is likely to occur where permeable surface soils lie over slowly permeable or impervious horizons at shallow depths.

5.18.3 Evaporation from the Soil

Evaporation is the process of vaporization by which liquid water is transferred into the atmosphere. The rate of evaporation of water depends on atmospheric temperature, soil humidity, organic matter content and wind velocity. Dry air has a greater capacity for absorbing moisture than moist air, hence evaporation increases under conditions of low humidity. High evaporation occurs at places of high temperature, low atmospheric humidity and high winds.

5.18.4 Transpiration by Plants

Transpiration is the discharge of water as vapour into the atmosphere by the leaves and stems of living plants. The amount and frequency with which irrigation water must be applied will be governed in part by the rate at which crops transpire. This rate of transpiration varies from practically nothing at young seedling stage during cold cloudy weather to a maximum during hot windy weather at a time when crops are growing most luxuriantly. Rapidly growing crops of a leafy nature may use about 5 cm of water per week during the peak growing season. The volume of water transpired by the plants depends mainly on the water at their disposal and also on temperature and humidity of the air, wind, intensity and duration of sunlight, stage of development of the plant, species of the plant and nature of the leaves.

5.18.5 Deep Percolations

When the amount of water entering the soil becomes greater than its water holding capacity, deep percolation occurs. This is an intensified form of infiltration in which a great loss of water takes place. Water goes beyond the exploration capacity of plant roots. This water moves due to the force of gravity.

5.19 USES OF WATER

There are many, many ways that we use our water, and that is partly why it is so important that we conserve our water. Water is our most precious resource. Water is vital to life. Humans, plants, and animals are made up of mostly water. All living things would die if it weren't for water. We use water for drinking, washing, cleaning, cooking, and growing our food as well as many, many other things. The average American uses around 150-250 gallons of water daily. Even more water is used by industries to generate electricity, manufacture things, and transport people and goods.

Common household uses consume a lot of water. It may take between 30 and 40 gallons for one bath. The average toilet uses about 5 gallons of water per flush. It takes 20-40 gallons of water for one shower. Washing machines use an average of 25 gallons per load. The kitchen sink takes roughly 20 gallons per day for preparing food and washing dishes. The bathroom sink, used for washing hands, shaving and brushing teeth, requires about 15 gallons per day. These numbers are estimated for the average household in America.

Much of our fresh water is also used outdoors for watering lawns, flower beds, and vegetable gardens, as well as washing cars and filling swimming pools. We must be careful not to pollute the water that we use outside. Many people use chemicals on their lawns and gardens and then water them with pure water. The water will wash the chemicals off of your plants and then run down a storm drain and go straight to the rivers and streams where fish make their homes. This kind of polluted water can kill fish and wildlife. We should always be mindful of whether or not we are polluting the water around us. We need to take care of our water, not abuse it.

Every town and city, whether small or big, uses water. Cities use water for fire fighting, street cleaning, and watering public areas such as parks, grass, trees, shrubs, and flowers.

Water is also used to fill public drinking fountains, including those at schools and libraries. All of the different businesses in your community also use water. Think about all the water that is used by restaurants, hospitals, laundries, dry cleaners, golf courses, hotels, car washes, beauty shops, barber shops, gas stations, and health clubs as well as all of the other businesses in town. These all add up to quite a big demand on Washington's water supply. We need to try to think about how many people need to use water and conserve our water so that there is enough to go around.

The amount of water needed to run a farm is tremendous. When we think of water on a farm, we think of watering crops; but the amount of water needed on a dairy farm is a huge amount also. Chickens, pigs, sheep, and all the other animals in a farmyard need drinking water to stay alive. Food must be grown for them to eat, and water is also required in the cooling systems used to keep production meat fresh. Vegetable and grain crops also require water. Water is used in spreading fertilizers, herbicides, and pesticides, which produce a greater crop yield, but can also contaminate the water. Most of the water used on farms is used for irrigation. Studies show that by using drip irrigation, farmers can conserve up to 60% of the water that it would normally take to irrigate their crops using other systems of irrigation.

It takes about 26 gallons of water to produce one ear of corn while it takes about 2000-2500 gallons of water to produce one pound of beef. Around 120 gallons of water is required to produce one egg. About 300 gallons of water is needed to produce one loaf of bread, and it takes about 12,000 gallons of water to grow a bushel of wheat. Believe it or not, about 1,400 gallons of water are used during the final production of one fast-food meal including a burger, fries, and a soft drink.

Hydroelectric plants are the largest users of water. Hydroelectric plants capture the kinetic energy of falling water to make electricity. They do this with a dam. The dam forces the water level to go up so that the water will have more power when falling. The force of the falling water pressing against the turbines' blades cause them to spin. The spinning turbines transmit the kinetic energy of the falling water to generators. The generators spin when the turbines spin generating electricity that will be transmitted on the power lines to homes and businesses.

Of all the electricity in the world, about 20% is generated by hydropower. About 10% of all the electricity in the United States is provided by hydropower. Hydropower generating prevents a lot of pollution. Hydropower generating is clean and does not leave any waste. Because of the electricity generated by hydropower, the amount of oil and coal needed to produce enough electricity is reduced. It prevents the need to burn about 22 billion gallons of oil or 120 million tons of coal each year.

The amount of electricity that a hydroelectric plant produces depends on two things: how far the water falls and the quantity of water falling. The higher the dam, the further the water falls and the more electric power produced. If the water falls twice as far, there will be twice as much electricity generated. The quantity of water that falls also affects the amount of power produced. The more water that flows through the turbines making them spin, the more electric power produced.

Water is also essential in industry. It is heated and the steam is used to run machinery. Water is used to cool hot metal such as in the production of steel. Water is also used to cool the air. It is an important element in many products like chemicals, drugs, lotions, shampoos, cosmetics, cleaners, and also beverages. Water is used in processing food and in innumerable factories and industrial processes including the manufacturing of paper. Water used in processing foods and beverages must be absolutely clean, while other industries such as a manufacturing plant may use a lower quality of water. "In the early 1900s, American industry used about 10 to 15 billion gallons of water a day. With the huge growth in industry following World War II, the industrial use of water also grew. By 1980, industry was using about 150-200 billion gallons each day." - Water: A Resource in Crisis by Eileen Lucas

Recreation is another way that we use and enjoy Washington state's water. Many people enjoy fishing, boating, sailing, canoeing, rafting, and swimming, as well as many other recreational activities that depend on water. Most of us enjoy at least one of these activities and would be disappointed if we could never do any of them again. If our water is continually polluted though, then it could become dangerous to go swimming and all of the fish could die so that no one could go fishing either. No one would enjoy boating, sailing, canoeing, or rafting in polluted water either. We need to remember that it is important to guard our waters from becoming polluted.

Transportation is yet another way that we use water. Many people use boats and ferries to commute to and from work every day. People also enjoy going on cruise ships or just going sailing. Water transportation has been a channel for growth in America, but that growth has not come without cost. The destruction caused by shipping includes fuel spills, which coat the water with oil and coal dust, and erosion of the shoreline which causes damage to the fish and other inhabitants of the water.

5.19.1 Role of People

It's 85 degrees and your throat is parched. As sweat drips down your neck, you envision a tall glass of cool water. You enter the kitchen, grab a glass, and turn on the faucet. Nothing happens. You stare, incredulous. Your mouth believes it's filled with cotton as you sprint to the bathroom, twist the faucet and wait for that nice clear water to rush out. Nothing happens. What's going on?? You stare at your empty glass and reevaluate your dog, who laps happily out of the toilet bowl.

Does that make you think twice about whether or not you should help conserve water? It is our responsibility to help conserve and preserve Washington's water so that we will always have plenty of pure, clean water to drink and use for other purposes as well.

There are many places in the world where there is not enough pure, clean water for the people, but people cannot live without drinking.

Every 8 seconds, a young person in a third world country dies from insufficient water supply or lack of pure water. Right now most of us have plenty of water and by just turning on the faucet, it is at our disposal. What we have plenty of with such little effort, we often take for granted. We don't usually think twice about the water that goes down our drain. When we turn on the faucet and water comes pouring out right down the drain, we don't even think about the fact that we are wasting water. The average American uses between 150-250 gallons of water daily. We could easily get by with a lot less if we just took the time to think about little things like turning off the faucet while brushing our teeth or shaving. We need to think when we use water, pay attention to what we are using it for, and we will soon realize just how much water we have been wasting.

The demands on our fresh water are ever increasing in today's modern society. New uses for this precious resource are being discovered constantly. Of all the water used in the United States, about 47% is used in agriculture, 44% inindustry, and only about 9% is used in households. Of the 9% that is used in households, only a very small portion is actually used as drinking water.

Americans use more water per household than any other country in the world. The results of the United States Geological Survey record that about 450 billion gallons of water are used in this country per day. That amounts to somewhere around 2,000 gallons per person per day. I know, you don't think that you use anywhere near that much water every day. You are right, you don't use that much water every day. That figure includes water that is used on farms and in factories to produce the food you eat; to cool your house, school, stores, and work places; to generate energy for running lights and appliances; and to produce manufactured goods. Even if you don't use 2,000 gallons of water every day, you probably use a lot more than you think. It is estimated that the average American uses between 100 and 200 gallons of water every day in the home.

People have always relied on lakes and rivers to provide food. Commercial fishing has been a prominent industry, as well as a vital source of food. Laws have been made about how many fish may be caught when and where so that people do not over-fish any body of water. When our water becomes polluted, many fish die and this creates a problem not only in the food chain, but also in the job line. Many people make their living fishing, and if there are not enough fish to keep all of the fisherman busy, then there could be many people without jobs.

The human body is made up of about 60-70% water. Your body uses water to digest food, maintain body temperature, remove wastes and toxins, cleanse the eyes, lubricate joints, and so much more. You probably drink a couple of glasses of water a day, but you also take in water from the food you eat. It is vital to your health that your body have enough water. You can probably live for several weeks without eating any food, but you would last only a few days without any water. With a loss of only 15% of your body's water, you are likely to die.

5.19.2 Fish and Wildlife

Animals rely on lakes and rivers for food and water. There are many animals that rely on the lakes and rivers and other bodies of water to supply them with life sustaining water, and for many of them food, also. Bears and birds are examples of animals that rely on fish as a main part of their diet. There are other animals that also eat a lot of fish. These animals would not know what to do if there were not enough fish for them to eat. Fish have always been a main part of their diet. Every part of nature depends on the other parts to help keep everything working right. Just as bears depend on salmon as a major part of their diet, and salmon depend on the water as their habitat, so every other thing in nature has things that it depends on. Some people call this the web of life, it is also known as an ecosystem.

Wetlands and Watersheds are also areas where numerable plants and animals other than fish make their home. When these areas are threatened by water pollution, the plants and animals are also threatened. Once a plant or animal becomes extinct, it is gone forever. Whales are one example of a species that has been in danger of becoming extinct. People have hunted whale for many years, and the water that they live in has also been polluted by oil and sewage. The ocean covers four-fifths of our planet. It is alive with activity - from tiny plants to gigantic whales. We must look after our water and preserve it, for without water there would be no life.

5.19.3 Nature

Plants are the source of all of our food. These plants receive energy from sunlight and store it in the food that they produce. This process called photosynthesis, is a process of the plant converting carbon dioxide and water to oxygen. This process is accelerated by a green pigment called clorophyll which is the pigment that makes the leaves green. The oxygen which is released during this process is what we breathe. This oxygen is also a life sustaining resource without which we could not live. Only green plants are capable of carrying out this amazing process of photosynthesis. The earth's entire source of food and most of its oxygen come from these plants.

Because plants are the source of all of our food and most of our oxygen, we need to make sure that they have enough water to carry out the process of making food and oxygen. Plants need water. They absorb the water and then let it pass through to their leaves where it is evaporated. Without water, there would be no life and there would be no life without plants either. It is necessary that we recognize the importance of plants and realize that we must conserve water so that we will continue to have a good supply of food and oxygen which we need to sustain life.

Forests provide a natural habitat for wildlife in Washington. Forests slow the water flowing down slopes, which reduces the amount of sediment runoff that contaminates rivers and streams. This keeps the water purer and helps give fish a healthier place to live. The forests also provide the earth with healthy topsoil. In the fall when the leaves and pine needles drop to the ground, they disintegrate and leave very healthy topsoil.

We need to protect the forests. They are an important part of our state. The trees purify the air that we breathe just as green plants do. Trees depend on water to live just as all other living things do. We should never cut down a tree within 100 feet of water. Doing this would contaminate the water and hurt the fish and other life in the water. When we cut down trees, we should plant new trees. This way there will always be trees available to meet our needs.

If we abuse our natural resources now, we will ruin them, and they will not be available for people in the future. This means that our children and grandchildren may not have enough clean, pure water to drink if we don't conserve it for them and ourselves. We have a responsibility to the future. We must protect and conserve our state's natural resource of water!

5.19.4 Pollution

Scientists have specified eight general categories for the major kinds of pollutants. These categories include sediment, sewage and organic wastes, plant nutrients, infectious agents, organic chemicals, salts and mineral substances, radioactive substances, and thermal (heat) pollution. We need to do our best to help keep these and other pollutants out of our streams, rivers, and lakes.

There are 8 categories for the major types of pollution: sediment, plant nutrients, infectious agents, organic chemicals, sewage and organic wastes, salts and mineral substances, and thermal (heat) pollution. It is important that we not only help conserve water, but that we help preserve water by trying not to pollute it.

Sediment runoff is when the loose, rich top soil erodes, and is washed into rivers, streams and lakes. Much of the sediment runoff comes from poorly managed land in rural areas. A good portion of sediment run off also comes from unprotected areas in or near cities. The solution to preventing water pollution by sediment runoff is to control erosion.

Plant nutrients is another category of water pollution that creates special problems. There are three main plant nutrients that cause problems: nitrogen, phosphorus, and potassium. The plant nutrients are the main problem mostly because of the large amounts of them that into surface waters.

Infectious Agents such as microorganisms and bacteria are especially significant pollutants of water because they cause diseases. The solution to plant nutrient pollution as well as infectious agent pollution is to properly treat all sewage and waste.

Organic chemicals include a wide variety of materials that pollute water including pesticides and herbicides used to get rid of undesirable plants and insects. These chemicals get into the water by being washed into rivers, streams, and lakes, and are usually carried in the sediment runoff. It is vital that we do as much as possible to keep these chemicals out of our rivers, streams, and lakes by disposing of the harmful products properly.

Sewage and Organic Wastes are extremely significant pollutants because of the quantity involved. In addition to the wastes being carried from homes, large amounts of organic wastes result from food processing and industrial operations.

Salts and Mineral Substances are pollutants that can be very troublesome. As water moves across the land, it picks up natural salts and minerals. Salt in water is a major problem especially in irrigation.

Mining waste has caused severe damage to some streams and rivers. There have been laws made to regulate mining which have helped to reduce the pollution of water from erosion and drainage of mines.

Thermal pollution is when the temperature of water is raised, like when water is used to heat or cool something. When water is heated and then put back into rivers, streams, and lakes, it causes a lot of problems for plants and animals that depend on that water. Plants and animals cannot survive for long when the temperature is much higher than normal. Some industries are using cooling towers and other things to help reduce the temperature of water they have used so that they can reuse it for cooling purposes.

These different kinds of pollutants are only a few of the many things that pollute water. You may have noticed that these pollutants are each related to the ways people use water. This means that if we are going to have clean, pure water we are all going to have to work together to see that our water is not damaged by the necessary ways we use it.

5.20 CANAL IRRIGATION SYSTEMS: WATER EFFICIENCY AND WATER PRODUCTIVITY

Much more challenging than that of groundwater is the improvement of performance of canal systems, which will remain the dominant water transfer technique in developing countries in the future. The standards used for design of irrigation canals in many countries have not kept up with the development of new technologies. Most traditional delivery systems have no or little flexibility built into them. Their control strategies and control structures are inadequate for good water management practices. Water deliveries at farm level are not compatible with the key requirements of modern agriculture. Thoughts on how to assess the performance of canal projects are evolving.

The engineering concept of efficiency at different levels of a canal system is important for planning, designing and even operating a canal irrigation system. However, water resources experts have criticised this concept which could be misleading. These experts argue that seepage losses contribute to the recharge of aquifers and losses from a project can be recovered downstream if the water quality has not deteriorated beyond certain limits. These arguments are valid when considering the development of water resources at basin level or even at broader levels when aquifers extend beyond basin boundaries. Advocates of water conservation measures, such as canal lining, then respond that water quality may decline in the recycling process, lands adjacent to canals may be affected by waterlogging and salinization, and an important volume of water can be lost from wasteland. Furthermore, farmers have to support the cost of groundwater pumping to which opponents can argue that farmers use groundwater more efficiently than canal water which is too often unreliable. The application of water conservation measures in irrigation can have negative or positive impact on third parties. The theoretical debate is endless.

Efficiency alone is not a sufficient indicator to define the performance of an irrigation system. Indicators of water productivity expressed in terms of production or value of production per unit of water are very important in increasingly water-scarce situation. A canal irrigation system may have high conveyance efficiency with a minimum of seepage and operational losses. However, if water delivery is too rigid or unreliable, there will be considerable waste further down at the farm level. A water productivity indicator provides a global indication of the effectiveness of water conservation measures and of the quality of service provided to the users, as well as the farm use of water and other inputs.

5.21 WATER CONSERVATION SUMMARY

In 1990, 30 states in the US reported 'water-stress' conditions. In 2000, the number of states reporting water-stress rose to 40. In 2009, the number rose to 45. There is a worsening trend in water supply nationwide. Taking measures at home to conserve water not only saves you money, it also is of benefit to the greater community.

Saving water at home does not require any significant cost outlay. Although there are water-saving appliances and water conservation systems such as rain barrels, drip irrigation and on-demand water heaters which are more expensive, the bulk of water saving methods can be achieved at little cost. For example, 75% of water used indoors is in the bathroom, and 25% of this is for the toilet. The average toilet uses 4 gallons per flush (gpf). You can invest in a ULF (ultra-low flush) toilet which will use only 2 gpf. But you can also install a simple tank bank, costing about \$2, which will save .8 gpf. This saves 40% of what you would save with the ULF toilet. Using simple methods like tank banks, low-flow showerheads and faucet aerators you can retrofit your home for under\$50.

By using water-saving features you can reduce your in-home water use by 35%. This means the average household, which uses 130,000 gallons per year, coulod save 44,00 gallons of water per year. On a daily basis, the average household, using 350 gallons per day, could save 125 gallons of water per day. The average individual, currently using 70 gallons per day, could save 25 gallons of water per day.

When buying low-flow aerators, be sure to read the label for the actual 'gpm' (gallons per minute) rating. Often, the big box retailers promote "low-flow" which are rated at 2.5 gpm, which is at the top of the low-flow spectrum. This may be needed for the kitchen sink, but we find that a 1.5 gpm aerator works fine for the bathroom sink and most water outlets, delivering the same spray force in a comfortable, soft stream. Finally, it should be noted that installing low-flow aerators, showerheads, tank banks and other water-saving devices usually is a very simple operation which can be done by the homeowner and does not even require the use of tools. Water conservation at home is one of the easiest measures to put in place, and saving water should become part of everday family practice.

5.22 WATER CONSERVATION SOLUTIONS: WATER IS LIFE

Because of these circumstances, resources of consumable water are very small. The closest estimate to date is that there is about 1 per cent of water on the earth that is suitable for drinking. This is serious, because in many ways water represents life. Albert Szent-Gyorgyi, the Hungarian biochemist and Nobel Prize winner for medicine says, "Water is life's matter and matrix, mother and medium. There is no life without water."

Consumable and fresh water is the need of every living being on earth. Humans, animals, other creatures, and agriculture all need water to survive and thrive. Without water our planet would become a vacant and barren wasteland.

And if you are thinking that water scarcity is just an issue in third world countries… think again. "Twenty-one percent of irrigation in the United States is achieved by pumping groundwater at rates that exceed the water's ability to recharge," wrote water expert Tony Clarke of the Polaris Institute.

Having water available in its clean and purest form is a blessing, and should be treated with the utmost respect. It is our obligation to save and conserve water as much as we can. In this way we will be helping ourselves and all of humanity.

Below are simple water conservation solutions. And while you practice these water conservation methods give a silent expression of thanks for having the clean water that so many others do not.

5.22.1 Water Conservation Solutions: Simple Things you can Do

- Take shorter showers.
- Close the tap at once after using the water.
- While brushing, only open the tap when you need water.
- Do not over-water plants.
- Wash your car less frequently and use smaller amounts of water, or use no water when you wash your car.
- Fix leaky pipes immediately upon discovery.

- Slow the flow of water to consume less water, using less energy.
- Operate the washing machine only when you have a full load of laundry.
- Don't throw towels in the laundry after only one use.
- Run your dishwasher only when you have a full load.

These are some ideas to conserve water. Now, let's look at what solutions are available on the market that could help you use less water.

5.22.2 Water Conservation Solutions: Products Available that Help the Cause

- Use the water efficient taps. These taps turn off automatically when not in use.
- Buy appliances that are marked with water efficiency labels. These energy efficient machines use about half the amount of the water as standard appliances.
- Use water efficient showerheads.
- Install a dual flush toilet.
- Install aerators, flow regulators, flow restrictors, and pressure-limiting valves to your taps. These measures will minimize the flow of water considerably.
- Purchase washing machines with front-load washing. These machines use much smaller amounts of water compared to the other machines.
- Keep your appliances at the recommended level to ensure greater efficiency.
- Don't use water to wash your car. Yes, you read that correctly. Find out about this innovative new product at Freedom Waterless and see how it works.

Consider each of these points as a thank you to the planet. When individuals begin to take part in creating a healthier environment, the law of cause and effect will start to spread in a life-enhancing way.

5.22.3 Water Conservation Devices: Dual Flush

The United States, like Australia and many other countries are experiencing more erratic weather that places a significant strain on water resources.

This can no longer just be ignored. Any house hold that still uses an old toilet flushing cistern, should give very serious thought to converting to dual flush. A dual flush system is probably the single most significant water saving invention to date. The Dual Flush toilet that was developed 25 years ago in Australia is such a success story that it has been mandatory throughout Australia for many years, and most of the country has been retrofitted with Dual Flush, significantly reducing the strain on their fresh water supply. They are now available around the world from different manufacturers.

5.22.4 Recycling Water

Water is a common substance that is essential to all known forms of life. Water covers 70 percent of the planet's surface. On the planet Earth water is mostly found in oceans and large bodies of water. 1.6 percent of the Earth's water is found underground in aquifers. A tiny percentage of the world's water is found in the bodies of plants and animals and in manufactured goods.

Humans and other animals need access to a supply of fresh and clean drinking water for their survival.

The availability of safe drinking water in almost all areas of the world has increased to a large extent over the last few decades.

Despite this fact some observers have interpolated that by 2025 a large part of the world's population will be facing shortages of water.

One household in the USA for example uses on average 127,000 gallons of water per year. But if greater use was made of low-flow fixtures on their taps, washing machine, shower head and toilet, one USA household could save 40,000 gallons of water per year and 30% on their water bill. If every household used low-flow fixtures, we could save 5.4 billion gallons/day which is worth $11.3 million dollars!

The United Nations calls this situation a water crisis. The role of water in the world's economy is enormous because it is used as a dilutant for a wide variety of chemical substances in industrial and medical applications.

Agriculture is the biggest consumer of freshwater using over 70 percent of freshwater supplies.

Recycling water is the use of treated and processed wastewater for useful purposes like agricultural irrigation, industrial processes, toilet flushing, and groundwater replenishment which is also called a groundwater recharge. Water recycling is also known as water reusing or water reclamation. These terms are often used interchangeably.

5.22.5 How can Water Recycling Benefit Us?

Water recycling provides enormous environmental benefits. It also provides an additional source of water for various purposes. This a list of some benefits that water recycling can present.

- Water recycling decreases the extraction of water form sources that may be dwindling and may stop being viable as habitats for valuable and endangered wildlife.
- Recycling wastewater can decrease the discharge of effluents that may damage and pollute the ecosystems of the sensitive bodies of water.
- Recycled water can be used to create new wetlands or to enhance and improve the quality of existing ones.
- Water recycling can reduce and prevent pollution by leaving damaging pollutants at the treatment plant.

5.22.6 Recycling Water: What it Does For the Future?

Recycling water has been established as an effective and safe way to provide non-potable water for a wide range of industrial, home, and agricultural use. The non-potable utilization of recycled water can only continue to grow. In numerous parts of the United States the uses of recycled water are increasing to accommodate the environmental situation and growing demands for water.

Technological advances in wastewater treatment and health studies in the use of indirect potable water have indicated the possible planned indirect potable reuse might soon become feasible. As the need for more water grows and environmental pressures increase, the necessity for water recycling can only increase as well. Water recycling will need to have a greater role in providing water for all the needs of society. We must work together to make water recycling and water conservation effective enough to help us sustain our valuable and vital water resources.

5.23 WATER CONSERVATION FACTS AND TIPS

- Less than 2% of the Earth's water supply is fresh water.
- Of all the earth's water, 97% is salt water found in oceans and seas.
- Only 1% of the earth's water is available for drinking water. Two percent is frozen.
- The human body is about 75% water.
- A person can survive about a month without food, but only 5 to 7 days without water.
- Every day in the United States, we drink about 110 million gallons of water.
- Landscaping accounts for about half the water Californians use at home. Showers account for another 18 percent, while toilets use about 20 percent.
- Showering and bathing are the largest indoor uses (27%) of water domestically.
- The average American uses 140-170 gallons of water per day.
- If every household in America had a faucet that dripped once each second, 928 million gallons of water a day would leak away.
- There are 7.48 gallons in a cubic foot of water. Therefore, 2000 cubic feet of water is 14,960 gallons.
- An acre foot of water is about 326,000 gallons. One-half acre foot is enough to meet the needs of a typical family for a year. There are 7.48 gallons in a cubic foot of water.
- It takes 3.3 acre feet of water to grow enough food for an average family for a year.
- A leaky faucet can waste 100 gallons a day.

- One flush of the toilet uses 6 ½ gallons of water.
- An average bath requires 37 gallons of water.
- An average family of four uses 881 gallons of water per week just by flushing the toilet.
- The average 5-minute shower takes 15-25 gallons of water--around 40 gallons are used in 10 minutes.
- Take short showers instead of baths.
- A full bathtub requires about 36 gallons of water.
- You use about 5 gallons of water if you leave the water running while brushing your teeth.
- If you water your grass and trees more heavily, but less often, this saves water and builds stronger roots.
- Each person needs to drink about 2 ½quarts (80 ounces) of water every day.
- Water your lawn only when it needs it. If you step on the grass and it springs back up when you move, it doesn't need water. If it stays flat, it does need water.
- Run your dishwasher and washing machine only when they are full.
- When washing a car, use soap and water from a bucket. Use a hose with a shut-off nozzle for rinsing.
- Never put water down the drain when there may be another use for it such as watering a plant or garden, or cleaning.
- Avoid flushing the toilet unnecessarily. Dispose of tissues, insects and other such waste in the trash rather than the toilet.
- When washing dishes by hand, fill one sink or basin with soapy water. Quickly rinse under a slow-moving stream from the faucet.
- An automatic dishwasher uses 9 to 12 gallons of water while hand washing dishes can use up to 20 gallons.
- Store drinking water in the refrigerator rather than letting the tap run every time you want a cool glass of water.
- Water lawns during the early morning hours, or evening when temperatures and wind speed are the lowest.
- This reduces losses from evaporation.
- Do not hose down your driveway or sidewalk.
- Use a broom to clean leaves and other debris from these areas.
- Using a hose to clean a driveway wastes hundreds of gallons of water.
- Don't leave the water running when brushing your teeth or shaving. Get in the habit of turning off the water when it's not being used.

- Use of bowl of water to clean fruits & vegetables rather than running water over them. You can reuse this for your house plants.
- Public water suppliers process 38 billion gallons of water per day for domestic and public use.
- Approximately 1 million miles of pipelines and aqueducts carry water in the U.S. & Canada. That's enough pipe to circle the earth 40 times.
- About 800,000 water wells are drilled each year in the United States for domestic, farming, commercial, and water testing purposes.
- More than 13 million households get their water from their own private wells and are responsible for treating and pumping the water themselves.
- Industries released 197 million pounds of toxic chemicals into waterways in 1990.
- You can refill an 8-oz glass of water approximately 15,000 times for the same cost as a six-pack of soda pop.
- A dairy cow must drink four gallons of water to produce one gallon of milk.
- 300 million gallons of water are needed to produce a single day's supply of U.S. newsprint.
- One inch of rainfall drops 7,000 gallons or nearly 30 tons of water on a 60' by 180' piece of land.

5.23.1 Facts on Saving Water

According to the United States Environmental Protection Agency (EPA), a family of four uses 400 gallons of water every day. Despite the majority of the Earth being covered in water, only 1 percent of it is useable - the rest is salty ocean water or permanently frozen. Water is a limited resource, and with the population continuing to grow, it is essential to practice water conservation. Here are some water conservation facts and ways to reduce the amount of water you use or waste.

Water Conservation Facts

Many of your everyday activities use water - from bathing and cleaning the house to drinking and cooking. On a broader scale, water is necessary for agriculture, power plants and industrial or mining. The American lifestyle requires a great deal of this precious resource. Here are some surprising water usage statistics from the EPA.

- The average American family spends $500 per year on its water and sewer bill.
- Treating and delivering water takes a lot of energy - letting your faucet run for 5 minutes takes as much energy as letting a 60-watt lightbulb run for 14 hours.
- More than 30 percent of household water is for outdoor use, such as watering lawns.

- Turning off the tap water while you brush your teeth can save 8 gallons of water a day - that equals about 200 gallons of water saved per month.
- A faucet that leaks 1 drip per second wastes 3,000 gallons of water each year.
- Leaky toilets can waste up to 200 gallons per day, the equivalent of flushing the toilet 50 times for no reason.
- Older toilets use 60 percent more water per flush as compared to newer high-efficiency model (look for the WaterSense label when buying a new toilet).
- The average bath requires 70 gallons of water while a shower uses 10 to 25 gallons.
- A running hose outdoors - for washing cars or watering plants - can waste 6 gallons of water per minute when left unattended.
- US power plants use 136 billion gallons of water each day.
- US agriculture uses 142 billion gallons of water per day for irrigation.
- US industrial facilities require more than 20 billion gallons of water every day.
- Lower water levels in fresh water reservoirs can contribute to concentrated levels of both natural and human-based pollutants in the water supply.

5.24 WATER CONSERVATION PRACTICES

Low-flow plumbing fixtures and retrofit programs are permanent, one-time conservation measures that can be implemented with little or no additional cost over the lifetime of the fixtures. In some cases, these fixtures can even save the residents money over the long term. The most commonly recommended low-flow plumbing fixtures are pressure reduction devices, faucet aerators, toilet displacement devices, low-flush toilets, low-flow showerheads, and plumbing modifications for gray water reuse.

Pressure Reduction: Homeowners can reduce the water pressure in a home by installing pressure reducing valves. A reduction in water pressure can save water in other ways: it can reduce the likelihood of leaking water pipes, leaking water heaters, and dripping faucets.

Faucet Aerators: Faucet aerators, which break the flowing water into fine droplets and entrain air while maintaining wetting effectiveness, are inexpensive devices that can be installed in sinks to reduce the volume of water used. Aerators are easily installed and can reduce the volume of water use at a faucet by as much as 60% while still maintaining a strong flow. More efficient kitchen and bathroom faucets that use only 7.5 l/min, in contrast to standard faucets, which use 12 to 20 l/min, are also available.

Toilet Displacement Devices: Non-toxic bricks or plastic containers (e.g., milk jugs filled with water or pebbles) can be placed in a toilet tank to reduce the amount of water used per flush. By placing between one and three such containers in the tank, more than 41 of water can be saved per flush. A toilet dam, which holds back a reservoir of water when the toilet is flushed, can also be used instead of the displacement device to save water.

Low-Flush Toilets: Conventional toilets use 15 to 201 of water per flush, but low-flush toilets use only 61 of water or less. Since low-flush toilets use less water, they also reduce the volume of wastewater produced. A schematic of a low-flush toilet is shown in Figure 43. Even in existing residences, replacement of conventional toilets with low-flush toilets is a practical and economical water-saving alternative.

Low-Flow Showerheads: Showers account for about 20% of the total indoor water use in an household. By replacing the standard 18 l/min showerheads with 10 l/min showerheads, which cost less than $5 each, a family of four can save approximately 80,000 l/year. Properly designed low-flow showerheads, currently available, are able to provide the quality of water delivery found in higher volume models.

Gray Water Use: Domestic wastewater composed of washwater from kitchen sinks and tubs, clothes washers, and laundry tubs is called gray water. Gray water can be used by homeowners for home gardening, lawn maintenance, landscaping, and other uses that do not require potable water. The level of contamination of gray waters is minimal; however, the plumbing modifications needed to make use of this water should not allow its contamination by wastes from the toilets, which have the potential to spread disease, cause undesirable odors, and result in aesthetic degradation of homestead yards and gardens.

1. The 6 liter flush design of this gravity toilet has a different flush mechanism.
2. Steep bowl sides and a narrow trapway to allow the siphoned water to gain velocity for more effective removal of waste.
3. This is where the water pushes waste into the trapway.
4. Stored water flows into the bowl.

5.24.1 Landscaping Water Conservation Practices

Drought-Tolerant Plants: Water conservation in landscaping can be accomplished through the use of plants that need little water, thereby saving not only water but labor and fertilizer as well. Careful landscape design can significantly reduce water use; it can also take advantage of native plants which have evolved water-saving or water-tolerant characteristics ideally suited for the local climatic conditions. Use of native plants can also help to minimize the spread of exotic plant species that disrupt local ecosystems. In addition to the selection of the plant species to be used in landscaping, practices such as the use of low precipitation rate sprinklers that have better distribution uniformity, bubbler/soaker systems, and/or drip or point irrigation systems can also conserve water used for landscaping purposes.

Xeriscaping: Xeriscaping is an innovative approach to landscaping that promotes water conservation and pollution prevention. Traditional landscapes might incorporate one or two principles of water conservation, but xeriscaping uses planning and design, soil analysis, selection of suitable plants, practical turf areas, efficient irrigation, use of mulches, and appropriate maintenance to create an appropriate landscape for a given climatic condition. Xeriscaping is most successfully practiced in arid and semi-arid areas, where it has proved useful for minimizing irrigation and external maintenance needs while presenting an attractive appearance.

5.24.2 Agricultural Water Conservation Practices

Water saving irrigation practices fall into three categories: field practices, management strategies, and system modifications. Examples of these practices include, respectively, the chisel plow aeration of extremely compacted soils, furrow diking to prevent uncontrolled runoff, and leveling of the land surface to distribute water more evenly.

Irrigation Scheduling: Improved irrigation scheduling can reduce the amount of water required to irrigate a crop effectively by reducing evaporative losses, supplying water when most needed by the irrigated plants, and applying the water in a manner best suited to the plants being irrigated. A careful choice of the rate and timing of irrigation can help farmers to maintain yields with less water. In making scheduling decisions, irrigators should consider:

- The uncertainty of rainfall and the timing of crop water demands.
- The limited water storage capacity of many irrigated soils.
- The finite pumping capacity of most irrigation systems.
- The price of water and changes in water prices as additional operators increase water demand.

Irrigation Management: Management strategies involve monitoring soil and water conditions and collecting information on water use and efficiency. The methods include measuring rainfall, determining soil moisture levels, monitoring pumping plant efficiency, and scheduling irrigation. Typical system modifications include the addition of drop tubes to a center pivot irrigation system, retrofitting wells with smaller pumps, installing a surge or demand irrigation system, and/or constructing a tailwater or return flow recovery system.

5.24.3 Industrial and Commercial Users Water Conservation Practices

Water recycling is the reuse of water for the same application for which it was originally used. Recycled water might require treatment before it can be reused. Cooling water recirculation and washwater recycling are the most widely used water recycling practices. The following guidelines should be used when considering water reuse and recycling in industrial and commercial applications:

- **Identification of water reuse opportunities:** Are there areas within the factory or in the production process that currently use water only once that would be amenable to reuse?

- **Determination of the minimum water quantity needed for the given use:** Are there areas within the factory or in the production process where more water is being supplied than is needed to accomplish the purpose?
- **Identification of wastewater sources that satisfy the water quality requirements:** Does the process require potable water or water of a lesser standard? Can the same result be achieved with lower-quality water?
- **Determination of how the water can be transported to the new use:** What modifications, if any, in the process or factory may be needed to permit recovery and recirculation/recycling of the water currently sent to waste? What additional treatment may be necessary to reuse this water? What is the relative cost of the required modifications versus the cost of the raw water over the life of the modifications?

Cooling Water Recirculation: Recycling water within a recirculating cooling system can greatly reduce water consumption by using the same water to perform several cooling operations. The water savings are generally sufficiently substantial to result in an overall cost saving to industry. Such savings can be even greater if the waste heat is used as a heat source elsewhere in the manufacturing process. Three cooling water conservation approaches are typically used to reduce water consumption: evaporative cooling, ozonation, and heat exchange.

Washwater Recycling: Another common use of water by industry is in the use of fresh or deionized water for removing contaminants from products and equipment. Deionized water can generally be recycled after its first use, although the reclamation treatment cost of recycling this water may be as great as or greater than the cost of purchasing raw water from a producer and treating it. The same processes required to produce deionized water from municipal water can be used to produce deionized water from used washwater. It is also possible to blend used washwater with raw water, which also would result in an overall water saving. The reuse of once-used deionized water for a different application within the same factory should also be considered as a water conservation option. For example, used washwater may be perfectly acceptable for washing vehicles or the factory premises.

5.24.4 Water Conservation Practices for Water Utilities

Common practices used by water supply utilities include metering, leak detection, repairing water lines, well capping, retrofitting programs, pricing, wastewater reuse, and developing public education programs and drought management plans.

Metering: The measurement of water use with a meter provides essential data for charging fees based on actual customer use. Submetering may also be used in multiple-unit operations such as apartment buildings, condominia, and mobile homes to indicate water use by individual units within a complex. In such cases, the entire complex of units might be metered by the main supplier, while the individual units might be monitored by either the owner or the water utility.

Leak Detection: It has been estimated that in many distribution systems up to half of the water supplied by the water treatment plant is lost to leakage; even more may be lost due to unauthorized abstraction. One way to detect leaks and identify unauthorized connections is to use listening equipment to survey the distribution system, identify leak sounds, and pinpoint the locations of hidden underground leaks. Metering can also be used to help detect leaks in a system. It is not unusual for unaccounted water losses to drop by up to 36% after the introduction of metering and leak detection programs.

Water Distribution Network Rehabilitation: A water utility can improve the management and rehabilitation of its water distribution network by a well-planned preventive maintenance program based on a sound knowledge of the distribution network. This knowledge is often embodied in a distribution system database that includes the following data:

- An inventory of the characteristics of the system components, including information on their location, size, and age and the construction material(s) used in the network.
- A record of regular inspections of the network, including an evaluation of the condition of mains and degree of corrosion (if any).
- An inventory of soil conditions and types, including the chemical characteristics of the soils.
- A record of the quality of the product water in the system.
- A record of any high or low pressure problems in the network.
- Operating records, such as of pump and valve operations, failures, or leaks, and of maintenance and rehabilitation costs.
- A file of customer complaints.
- Metering data.

Through the monitoring of such records, advance warning of possible problems can be achieved. For example, excessive water use, or numerous complaints or demands for spare parts, could be early warning signs of an impending breakdown in the system. This system should also include a regular program of preventive maintenance to minimize the possibility of system failures.

Well Capping: Well capping is the sealing of abandoned wells. In the case of artesian wells, rusted casings can spill water in a constant flow into drainage ditches, resulting in evaporative loss or runoff losses. In non-artesian wells, uncapped abandoned wells form points of entry for contaminants into the groundwater system.

Pricing: Placing an economic value on freshwater is the principal means of achieving water conservation. Pricing provides a financial incentive to conserve water. Rate structures may be variable and/or graduated, with prices fixed on the basis of class of service (residential versus industrial or agricultural, for example) and quantity used (for example, the unit price for quantities below 400 l/day might be significantly lower than for quantities which exceed that amount for a single-family residence). Pricing has the advantage of minimizing the costs of overt regulation, restrictions, and policing, while providing a high degree of freedom of choice for individual water customers.

Retrofit Programs: Retrofitting involves the replacement of existing plumbing fixtures with equipment that uses less water. The most successful water-saving fixtures are those which operate in the same manner as the fixtures being replaced; for example, toilet tank inserts, faucet aerators, and low flow showerheads do not significantly change the operation of the systems into or onto which they are placed, but they do result in substantial water savings.

Water Audit Programs: Various types of audits can be undertaken. For example, residential water audits may involve sending trained water auditors into participating households, free of charge, to encourage water conservation efforts, or providing them with record sheets to note down their water use for external analysis. Water audits may also be undertaken in commercial and industrial facilities, and may be combined with an assessment of the potential for implementing water reuse and recycling programs. A pre-implementation and post-implementation water audit in factories adopting a reuse and recycling program would be a valuable means of demonstrating and quantifying the water savings achieved.

Public Information and Education: Public information and education programs can be undertaken to inform the public about the basics of water use and conservation. Programs should be developed for specific applications and may be targeted at specific user groups or age groups; for example, at housekeepers, to encourage domestic water conservation, or at schoolchildren, to provide information on the wider implications of water conservation for future consumption, the environment and other uses. Basic information should include the following:

- How water is delivered and how wastewater is disposed of.
- The costs of water and water supply services.
- Why water conservation is important.

The programs should provide guidance on how the user groups and individuals can participate in conservation efforts. It should be noted that there is a large body of public information and education materials available, particularly in the United States, which may be obtained from a variety of public agencies and NGOs at little or no cost and form the basis of a local public awareness initiative.

5.24.5 Drought Management

Given the vagaries of the modem climate, in this period of climate change, it seems that droughts may be more severe or extensive than is the past. Many water conservation projects constructed to alleviate drought-induced water shortages are themselves victims of drought. Whether this may simply reflect changes in land use within a watershed that allow less water to infiltrate into the groundwater system, or results from population growth, which places greater demands on finite water resources, is not clear and rarely proved. In any case, many communities are currently experiencing a need to have drought management plans in place to ensure the greatest possible availability of freshwater during periods of below average rainfall.

Drought Management Planning: When rainfall is less than usual, there is less water to maintain normal soil moisture levels, stream flows, and reservoir levels and to recharge groundwater. Because of these varied sources and the multiple demands placed upon freshwater resources, a drought management plan should address a range of issues, from political and technical matters to public involvement. Some of the components of a typical drought management plan include the following:

- Identification of the available water resources.
- Tabulation of the multiple sectoral demands for freshwater.
- Description of possible shortfalls between supply and demand.
- Definition of the management measures required for various eventualities, and an agreed allocation schedule in the event that water rationing becomes necessary.
- Provision for user and public involvement in the drought management program.
- Promulgation of legislation, agreements, rules, and procedures to ensure a timely and equitable response to the onset of drought conditions.
- Issuance of a drought management event plan and public information materials to make it known.

5.25 EXTENT OF USE

Water conservation measures have been practiced primarily in the United States, although some Latin American countries have implemented specific measures. For example, in Brazil, the pharmaceutical, food processing, and dairy industries were required to pay effluent charges that contributed to reductions in water use and wastewater production of between 42% and 62%. In Mexico, increased water prices contributed to an increase in wastewater reuse and the recycling of cooling water.

Chile is the only country in the region with a comprehensive water law that has encouraged the development of water markets. The 1981 National Water Law established secure, tradable, and transferable water use rights for both surface and groundwaters. As a result, during periods of low rainfall, farmers shift from the production of water-intensive crops, such as corn and oilseeds, to higher-valued and less water-intensive crops, such as fruits and vegetables.

Water recycling is used at a Container Corporation of America Mill in Santa Clara, California (U.S.A), that manufactures paperboard from the recycled fibers of newspapers, corrugated cardboard clippings, and ledger paper. Historically, water has been used in this process for a variety of purposes. In recent years, however, the mill has begun recycling water used in its rinsing processes after clarification. The mill has also installed a closed loop cooling tower, which has resulted in an additional reduction in water use. These water conservation and use efficiency practices have resulted in an estimated saving of approximately 2.8 million l/day, compared to its 1980 water use rates. These water reductions amount to approximately 900 million l/year and saved the company approximately $348 200 per year.

5.25.1 Operation and Maintenance

Given the variety of measures that might be undertaken to address conservation needs within a specific geographic area, of which a number are mechanical but many may be technological or informational, it is difficult to identify specific operational requirements. However, some of the more obvious requirements include the following: low-flow water conservation devices require periodic maintenance and repair; leak detection equipment and meters require periodic testing and repair; drought and water conservation management strategies, such as pricing and user charges, require monitoring and enforcement; and well-capping programs require monitoring and trained personnel in order to be effective. Maintenance requirements range from regular inspections of mechanical devices to the review of legislation and conservation plans to ensure their continued relevance.

5.25.2 Level of Involvement

The installation and maintenance of low-flow household and irrigation devices may require governmental incentives in order to be accepted. In some cases, employees of the water utility may install and maintain these systems at little or no charge in order to effect the desired water savings. Alternatively, government regulations may be necessary to provide incentives for the implementation of industrial and agricultural water conservation measures. Government action is required in the promulgation of plumbing codes for new construction that will contribute to the adoption of residential water conservation measures. Government or utility involvement is also needed for leakage detection and the repair of distribution systems. Metering, in addition to requiring technical personnel and equipment to be effective, generally requires governmental action to implement and government authority to establish or regulate water tariffs. However, community participation and voluntary conservation are a key element if this technology is to be effective.

5.25.3 Effectiveness of the Technology

Water conservation measures are highly effective. However, this technology may not be too popular with consumers, who may be asked to pay a higher price for the water they consume, and can be, politically, very unpalatable. Nevertheless, studies carried out in Seattle, Washington, U.S.A., reported the following results from water conservation measures:

- According to detailed data on the performance of low-flow water devices in 308 single-family residences, indoor per capita water use dropped 6.4% after low-flow showerheads were installed.
- Easily installed aerators reduced water use at a faucet by as much as 60% while still maintaining a strong flow.
- A reduction in water pressure from 100 pounds per square inch (psi) to 50 psi at an outlet resulted in a water flow reduction of about one-third of the pre-existing use.
- Gray water reuse saved a volume of water equivalent to that needed to supply more than 7 000 residences and businesses.
- Outdoor water use was reduced by restricting watering times to the early morning or late evening; watering on cooler days, when possible, also reduced outdoor water use. All these measures contributed to reduced evaporative losses.
- As many as 600 l of water were saved when washing a car by turning the hose off between rinses; additional benefits and water savings were achieved by washing the car on the lawn, which both watered the lawn and reduced runoff.
- Sweeping sidewalks and driveways, instead of hosing them down, saved about 200 l of water every 5 minutes.

In other studies, such as an industrial water conservation project in California, the conversion of an industrial process from a single-pass freshwater cooling system to a closed-loop cooling system, with circulating chilled water, has saved an estimated 20 000 to 28 000 l/day, while cities in the hemisphere that have large, old, deteriorating systems, leak detection programs have been especially efficient in minimizing water losses.

5.25.4 Suitability

Water conservation measures are suitable and recommended for all public water supply systems, industries with high water use, agricultural enterprises, and individual residential users in Latin American countries and the Caribbean islands.

5.26 RESIDENTIAL WATER AUDIT PROGRAMS

Residential water audit programs involve sending trained water auditors to participating family homes, free of charge, to encourage water conservation efforts. Auditors visit participating homes to identify water conservation opportunities, such as repairing leaks and low-flow plumbing, and to recommend changes in water use practices to reduce home water use. The audit programs try to stretch existing water supplies by getting water users to use water more efficiently. The largest percentage of indoor use comes from bathing and toilet flushing. Therefore, the bathroom is an ideal place for water system operators to focus water conservation efforts.

5.26.1 Public Education

Public education programs can be used to inform the public about the basics of water use efficiency:

- How water is delivered to them
- The costs of water service
- Why water conservation is important
- How they can participate in conservation efforts

Public education is an essential component of a successful water conservation program. A number of tools can be used to educate the public [BROKEN]: bill inserts, feature articles and announcements in the news media, workshops, booklets, posters and bumper stickers, and the distribution of water-saving devices. Public school education is also an important means for instilling water conservation awareness (Grisham and Fleming, 1989).

Another way to provide public information and education, as well as to collect real-world data on water conservation and use efficiency, is through the use of demonstration projects. In Tucson, Arizona, the Casa del Agua, a single-family home, has been used to demonstrate and study water conservation and reuse techniques and technologies. In 1985, the University of Arizona designed and retrofitted the Casa del Agua with water-conserving fixtures, a rainwater harvesting system, gray water reuse and storage systems, and drought-tolerant plants. Measurements of water use and water quality at the Casa del Agua have provided a useful collection of data for evaluating the possible benefits of conservation techniques and technologies in a residential home.

A study of water demand in the United States using American Water Works Association (AWWA) data indicated that water users are more sensitive to a change in price in the South and the West than in the other regions of the country. Public education appears to have reduced water usage in the West. A heightened awareness of water's scarcity might make educational programs more effective in the West than in the rest of the country.

6

WATER TREATMENT METHODS

6.1 OBJECTIVES

After studying this chapter you will be able to understand:

- Filtration
- Ultraviolet Light
- Standard Practice for Household Use.
- Conal Irrigation Systems
- Water Recycling Process
- Ways to Conserve Water in the Home and Yard.

6.2 INTRODUCTION

Treatment of the water after the storage and before use is critical for both health of the users and maintenance of the system. The level of treatment will depend of the intended use of the water. For example, water used for landscape drip irrigation would not need the same level of treatment as water used for potable indoor purposes. In any case, we recommend a multi-barrier approach where more than one method of treatment are used to maximize effectiveness. For example, a common treatment-train for potable water in rural Texas is a set of cartridge filters followed by a ultraviolet light. Below are the different types of water treatment available.

6.3 FILTRATION

Filtration is similar to screening but on a smaller scale. There are various levels of filtration, and they are measured on a "micron" level. This measures the diameter of a particle that would be blocked by the filter. For example, a 5-micron filter would block particles in the water that are 5 microns or larger. (a micron is one-millionth of a meter). Filters can remove microorganisms, sediment, metals, and other organic matter. If larger sized filters are used, small microorganisms, such as bacteria can pass through, so a disinfection method is needed (explained next). It is important that they are checked and changed on a regular basis so that they maintain their effectiveness.

The goal of disinfection is to destroy the microbiological organisms that have the potential to cause illness or harm. Some of the smaller microorganisms, such as bacteria, may pass through large cartridge filters so they must be targeted with disinfection. Three disinfection methods common to RWH systems are chlorination, Ultraviolet Light (UV), and ozonation.

Chlorination uses the chemical of chlorine in either dry, liquid, or gas to kill microorganisms. It is very effective with viruses and bacteria. An advantage to chlorine is that is leaves a residual so that your entire distribution system remains disinfected. However, using chlorine can be dangerous, so it's important to follow manufacturer's guidelines. Only use chlorine that is approved for potable water use (standard ANSI/NSF 60). For more information see this guidebook by the TCEQ.

UV light is a common disinfection method used by Texas residents. The device works by sending water through a tube that contains the UV bulb (see picture on left). The light disrupts the DNA of microorganisms so that they can't reproduce. There are a Class A and Class B UV light. The Class A light is rated to destroy pathogens and is required to have an alarm to alert users if the device is not working. A Class B light cannot be used as a sole method of disinfection but helps to polish water that has already been treated. An important point to keep in mind is that UV lights are most effective when the water is clear. Any sediment in the water can block the pathogens from the light. This is why have the filtration BEFORE the UV light is so critical. The UV light is harmful to look at directly, so use approved safety glasses when maintaining the device.

Ozonation disinfects by introducing ozone gas to the water. It is usually done at the point where water is used in the distribution system or in the storage tank. It is a colorless gas that disinfects, oxidizes, deodorizes, and decolorizes. Ozone gas is toxic and installation and maintenance of this type of system must be done by a licensed professional.

Adsorption uses activated carbon (charcoal filters) to bind to harmful contaminants, such as VOCs. These are the types of filters commonly used on faucets. They can also be used to remove chlorine from water used for disinfection.

When I was growing up, I remember drinking out of a rain barrel with a ladle. My great aunt would yell out the door, "Remember not to drink off the top!" That was then and now is now.

A lot has changed in 4 decades. There are a lot more pollutants, and we are more aware of the risks. We now know that E. coli and other harmful bacteria can be passed along in untreated contaminated water. A report by Peter H. Gleick estimates that if no action is taken to address unmet basic human needs for water, as many as 135 million people will die from water-related diseases by 2020.

Rainwater harvesting is viewed by many, including the EPA, as a partial solution to the problems posed by water scarcity: droughts and desertification, erosion from runoff, over-reliance on depleted aquifers, and the costs of new irrigation, diversion, and water treatment facilities.

Harvested rainwater in the U.S. is used mostly for irrigation; however, there is a growing interest in using rainwater for drinking and other indoor uses. Over 50% of household water is used indoors; bringing rain indoors could save the expense and environmental costs of treating and transporting water.

Can rainwater be made safe to drink? Yes. How safe? As safe as your well or tap water. How do you make it safe for indoor use? By filtering and purifying it.

Contaminants in water may include algae, air pollution, bird excrement, and leaves, sand, and dust. Local wells have dealt with these problems for decades. Installation of filtration and purification equipment can remove these contaminants at home as well.

First, take measures to keep foreign matter out of the incoming rainwater. First flush devices, gutter screens and other screening mechanisms keep the rainwater as clean as possible before it enters the conveyance system. Using screens and filters will greatly reduce maintenance and lengthen the life of the pump and filtration/purification system.

Even the best screening systems will allow unwanted particulates into the cistern. To keep sediment where it belongs, at the bottom of your tank, screen incoming rainwater, give the remaining sediment time to settle, avoid disturbing it, and don't pull water from the bottom of the tank. Use a floating filter, which extracts water from the middle of the tank, leaving sediment undisturbed.

Next is filtration, which removes debris from the water. Disinfection or purification follows, which kills contaminants and removes harmful substances that may be present.

To determine what type of system you need, test the rainwater at a reliable laboratory. Without testing, you could spend a lot of money on equipment that will not give you safe water.

Filtration is included in every system, even simple irrigation systems. Examples of filtration systems include: screen filters, paper filters, and carbon or charcoal filters.

Almost all systems use multiple filters. For example, after gutter screens and/or a first flush device, a system often includes two in-line filters of increasing fineness, a carbon filter and a UV light. Each of these are described below to assist you in evaluating what might be the right alternative for your planned water use and required water quality.

In starting to evaluate filter options, it is imperative to know exactly what the filter system you select will actually remove from the water. National Sanitation Foundation/

American National Standards Institutions (NSF/ANSI) standards are the best, most stringent in the industry. Almost all water-filtration products are certified under NSF Standard 61 for Drinking Water System Components (see Related Topics). But the critical standards for contaminant removal are Standard 42, "Drinking Water Treatment Units - Aesthetic Effects," and Standard 53, "Drinking Water Treatment Units - Health Effects."

Standard 42 covers specific aesthetic contaminants (chlorine taste and odor, and visible particulates). Standard 53 covers health-related contaminants, such as Cryptosporidium, Giardia, lead, and volatile organic chemicals that may be present in drinking water. Systems that meet both of these standards are available, but expensive. Fortunately, the NSF website (see Related Topics) provides an easy way to search for units made by a specific manufacturer or that remove a specific contaminant.

6.3.1 Filters and Disinfection

Filters are measured in microns. One micron is about 1/25,000th of an inch. For comparison, sand is about 100 - 1,000 microns, a human hair is about 100 microns, a particle of dust is about 1 micron and a virus can be smaller than .01 micron.

The first filters in a system are cartridge filters. They range widely in what they are capable of removing and are used in a series (e.g., a 20 micron followed immediately by a 5 micron filter).

Filters are rated by the smallest size of particle they are capable of filtering. The smaller the micron size the better the filter. However, the finer the filter, the higher its cost and the slower its process. Filters have to be changed regularly, as an old, used filter is an excellent environment for microorganisms and potentially harmful pathogens.

For wells and rainwater systems a larger (e.g., a 50 micron) filter or equivalent screen (e.g., 300 mesh) should be used first to eliminate sand and large particles. This screen should be easily accessible and cleaned quarterly. Next is a 20 or 10 micron filter, followed immediately by a 10 or 5 micron filter. These are cleaned less frequently, but at least annually.

Filters will not eliminate all substances in the water. To create drinking quality water, filtration is always followed by disinfection. The EPA requires surface and ground water to be disinfected before it is consumed. Consequently, public water systems add disinfectants to destroy microorganisms that can cause disease in people and animals.

This is also necessary for rainwater, as the natural environment contains many microorganisms. Most are not harmful to us. Some, however, such as Giardia lamblia, can be deadly. These need to be eliminated from water before it is consumed.

Kinds of disinfection include chlorinization, ozonization, ultraviolet (UV) light, and membrane filtration. In evaluating disinfection methods, be aware that some actually create unhealthy byproducts that need to be treated.

The effectiveness of disinfection is judged by looking for an indicator organism that, if present, indicates other more harmful pathogens may be present. In getting a water test, this indicator organism is Total Coliform Bacteria that, if present, indicates other pathogens may be present as well.

Chlorine has been used as a disinfectant in public water systems for most of the past century. The introduction of chlorine to disinfect water has virtually eliminated waterborne diseases such as cholera, typhoid, dysentery and hepatitis, saving thousands of lives. However, it is often maligned due to suspected side effects.

For disinfection purposes, 2.3 fluid ounces of household bleach must be added per 1,000 gallons of water. Chlorine dosage rate will vary depending on quantity of water to be treated, pH and temperature.

A major downside of chlorine is that it is very reactive and easily combines with naturally occurring organic material to create harmful trihalomethanes (THMs) like chloroform. Chloroform is formed when chlorine reacts with either humic and/or fulvic acids, which are commonly found in water.

Because chlorine is reactive, it quickly dissipates. Keeping the dosage rate correct is critical when using this method of disinfection. THMs should be tested for in the water source if you are going to use Chlorine.

To reduce the possibility of harmful byproducts with the use of Chlorine, do the following:

Remove the byproducts after they have been created. This is costly, typically meaning other purification systems must be employed (e.g., Reverse Osmosis or other purifcation systems) or

The concentration of particulates/organics in the water before it is treated. This is accomplished by using filters to remove these substances from the water prior to chlorine treatment.

The Chlorine smell and taste can be removed with an activated carbon filter, often referred to as a charcoal filter. Granulated activated carbon filters are sometimes made from coconut shells and can be considered a "green" solution. Carbon block filters are compressed activated carbon, fused with a binding substance into a solid block.

6.3.2 Membrane Filtration

Membrane filtration is another alternative. Membrane filtration involves pushing water through a layer of material. Pressure-driven membrane technologies include microfiltration, ultrafiltration, nanofiltration and reverse osmosis. It is one of the few technologies capable of removing pharmaceuticals, and creates no byproducts.

Membrane technologies are more costly than other alternatives, but prices are rapidly declining. Most water purification experts expect membrane technology to become the prevalent technology in smaller systems over time as their price drops.

Choosing the right membrane technology is not straightforward, as the technology is changing and there are no real standards. Make sure you know what you need and match it to the type of system you are evaluating. Again, it is critical to test your water to know what you need before evaluating options.

Microfiltration (MF) is a membrane separation process using a pore size of .03 to 10 microns. Although this does not sound like a big range, when it comes to water purification, it is. The smaller the pore size, the more the system will remove. Microfiltration membranes are good for the removal of sand, silt, clay, algae, cysts and some bacteria.

Ultrafiltration (UF) is a membrane separation process using a pore size of approximately .002 to .1 microns. UF will remove all materials removed by an MF system, plus some viruses.

Nanofiltration membranes (NF) have an approximate pore size of only .001 microns. These small pore sizes require much more power to push water through the membrane and generate more waste than either MF or UF filtration systems. These systems eliminate virtually all cysts, bacteria, viruses, and other materials, including minerals. Consequently, the resulting water has a low pH that can be corrosive and needs to be remineralized, commonly using limestone, to raise the pH. Due to the greater power requirements, NF has yet to become mainstream.

Reverse Osmosis (RO) is the most widely used membrane technology today. Under counter RO systemThese systems remove particles as fine as .001 microns, are compact, simple to operate and have been in use for over a decade. RO systems remove radium, natural organics, pesticides, cysts, bacteria and viruses. To ensure contaminant reduction, seek out units certified by NSF for contaminant reduction and not just safety. RO systems produce waste water that needs to be processed; however, the newer units are becoming "greener," producing less, but still significant, waste. These units vary greatly in their efficiency, so make sure to ask about waste and efficiency when shopping for an RO system.

RO waste water contains a high concentration of the contaminants removed from the water, so dealing with this waste must be planned for when installing an RO system. Options for dealing with this water include plumbing through a greywater system to the irrigation system or directly to the septic system.

RO systems come in small under-the-counter units or whole-house systems. Prices will vary greatly for these units and only NSF-certified units should be considered. Under-the-counter units generally include a sediment filter, a carbon filter, the RO membrane and another carbon filter, and will generally cost under $1,000. A whole-house unit contains all the same components, but is capable of handling much larger water flow rates, and generally includes a calcite or equivalent filter to reduce the pH of the water, and a large storage tank (e.g., 20 – 50 gallons). The cost of a whole-house unit can run upwards of $8,000, depending on size of the house and family.

Regardless of system size, maintenance needs to be performed regularly. The most frequent maintenance is changing cartridges. Filters are used to protect the RO membrane from particle fouling. As these filters trap particles from the water supply, a reduction in pressure occurs. Many RO units include a low-pressure switch that prevents the RO from running if the pressure drops too low. Check the allowable pressure drop across the cartridge and compare this to the incoming feed pressure. If it is lower than manufacturer recommendations, the filters need to be replaced.

6.3.3 Distillation

The last commonly available purification technology is distillation. Distillation separates the water from the impurities through heating and then collecting the condensation. It is very energy intensive and loses about 5-10% of the water due to evaporation. Distillation removes almost all substances from the water with the exception of volatile organic chemicals (VOCs) that evaporate easily. To this end, some distillation systems are also equipped with carbon filters to remove the VOCs.

Distillation works slowly to reduce energy requirements and, like RO systems, will store the purified water in a tank for later use. In addition to using a lot of electricity to operate, distillation systems generate heat.

Distillation units producing 5 -12 gallons of water a day will typically cost about $1,500 - $2,000. Cost will increase as capacity increases and as options are added. High-end automatic home units with larger storage capacity may cost upwards of $4,000. New solar distillers give you the option of reducing the electrical requirements.

6.4 ULTRAVIOLET LIGHT

An alternative for disinfecting water is Ultraviolet (UV) light. UV lights have been used for nearly a century in Europe and are now common in the US. With UV lights, the water must always pass through a filtration system first. If no filter is used, pathogens and bacteria will cast shadows in the flowing water, thereby allowing live organisms to pass through unharmed.

UV light works by penetrating an organism's cell walls and disrupting the cell's genetic makeup, making it impossible to reproduce and rendering it harmless. Often it is claimed that it kills the microorganism, but it doesn't - it just makes them unable to reproduce and thus harmless. UV lights do not change the chemical composition of the water and leave behind no by-products.

For UV to be effective the right light dose must be used to a specific unit of water and the water must be clear of suspended solids and other particulates. Most UV units are usually insensitive to temperature and pH differences in the water, but manufacturers' fine print should be read and followed.

There are several issues with UV lights should be taken into consideration:

Replace the bulb at the manufacturer's specified intervals – generally after 9,000 hours, or about every 12 months;

UV light is not visible to the human eye, so it may appear to be lit and in fact is not working;

The glass enclosure around the light needs to be cleaned occasionally for the UV light to be effective;

If no backup light is installed the water needs to be shut off upstream of the bulb prior to replacing the unit. Generally it is prudent to disinfect the water downstream after the system has been shut down for any reason.

UV systemCorrect UV treatment is effective in reducing harmful pathogens from the water. It is generally recommended that home units include alarms to notify the user when a bulb needs to be serviced or the unit is not working. Purchase a unit that has an automatic bulb cleaner, to reduce maintenance requirements. Two units should be installed, so when one unit needs servicing the second unit can be turned on so there is no disruption in disinfecting the drinking water.

UV light manufacturers rate their systems to a given dosage at a given flow rate (e.g., 10 gallons per minute). When installing a UV light, make sure the flow rate of the UV unit is matched to your flow rate of water (i.e., the pump flow rate). If the pump rate is greater than that of the UV light, install a pressure regulator or flow restrictor.

To properly treat the rainwater, it must contain particulates no larger than 50 microns and contain no tannins, sulfur or sulfur-related bacteria, have less than 0.3 parts per million of iron, and less than 0.005 parts per million of manganese. Knowing whether these are in the water and need to be treated is a great reason to test your water before installing a system. If any of the above is present in the water, the filters must deal with these elements before the water is treated by a UV light. Most of these will not be present in rainwater, but could result from local air pollution or contamination of the conveyance system. Don't assume anything until your water has been lab tested.

The UV light unit is typically installed after all filtration and the resulting water is clean, bug-free and ready to use. Entry-level units will handle about 10 gallons per minute. The price of the unit will increase as options and flow rates increase.

6.5 STANDARD PRACTICE FOR HOUSEHOLD USE

A common practice in off the grid homes is to filter all the incoming rainwater and then store it in a small pressure tank. From the pressure tank the outgoing water is split into two separate paths - one path for potable and the other for non-potable water. A purification process is added to produce potable water. The major advantage of this approach is that it requires a much smaller unit and costs less, since it treats less water than a whole-house unit. But the disadvantage is that it requires a dual plumbing system – one to supply filtered but non-potable water to the toilets, clothes washer, irrigation faucets, etc., and one to supply potable water to the faucets.

An apparently low-cost, entry-level system is a countertop or pitcher type unit for potable water. However, when measured on gallons of water processed between changing filters, these units tend to be much more expensive in the long run. For example, a typical faucet unit available at most large hardware stores needs its filter changed every 100 gallons. For a family, this would be more than once a month and each filter costs about $30. This could cost nearly $500 a year, just for filters!

Before investing in filtration or purification equipment, invest in removing particulates before they enter into the system by installing gutter screens, leaf screens and roof washers. Removing materials before they enter the system is far easier and less expensive than dealing with them afterwards.

There is no perfect solution for disinfecting water, as all solutions have some environmental cost. Some require substantial energy, some create harmful by-products and some waste water. To save money, test your water (have you heard that before?) and get the right unit to solve your specific problem. Generally, the smaller the capacity the less expensive the unit will be overall, so get only what you need.

Lastly, remember that as the owner of a water system, it is your responsibility to maintain it. When you pay for utility-purified and -delivered water, maintenance is included in your bill. But when you own your water system, it is your responsibility to maintain it on a regular basis.

Rainwater can be safely used outdoors and indoors if the correct steps are taken to handle, store and clean it. Although not yet common in the US, indoor use of rainwater is practiced worldwide. As population growth continues, water rates increase and the desire to be "more green" and self-reliant increases, rainwater use will become more common here in the United States.

6.6 THE ADOPTION OF WATER CONSERVATION APPLICATION TECHNIQUES

A number of farm application irrigation technologies that could contribute to the conservation of water in both surface and piped systems have been developed. These techniques are generally adopted for the cultivation of high value crops. About 20 million hectares worldwide are now irrigated by sprinkler irrigation. Micro-irrigation as a technology has matured into reliable water and fertigation management system for crop production over the past decades and its usage continues to increase rapidly. A total of 3.2

million hectares are being irrigated by micro-irrigation techniques throughout the world, which represent one percent of the total areas irrigated in the world. A growth of more than six times over the last 20 years has been achieved, which is quite a success. Micro-irrigation is an important tool in the drive towards food security. Micro-irrigation systems have great potential to create significant opportunities for smallholder agriculture.

During the 1980s micro-irrigation started penetrating developing countries, mainly India. A few years later, fearing a water and food shortage, the Chinese Government declared drip irrigation technology to be amongst the top priorities for development in the last decade. However, Vidal et al., (2001) have shown that micro-irrigation "is not a miracle technology: excellent as well as poor results were obtained from these technologies, and their adoption really depends on farmers capacity to finance and operate them, as well as on the type of crop production".

According to the ICID database on modern irrigation techniques, it appears that the use of pressurized water application techniques is widely different between countries. The percentage of modern technique decreases with the total area irrigated in developed countries from a low of 21 percent in the USA and 16 percent in Italy to 90 percent or over in Northern European countries (France and Germany). Sprinkler has a much greater application compared to that of the drip systems with the exception of Cyprus, Israel and Jordan.

In South Asia and Africa low cost bucket-and-drip sets are becoming increasingly popular with farmers. In areas where shallow groundwater is plentiful, thousands of poor farmers in Bangladesh have used low-cost treadle pumps to supply water for crops for their own food security and additional income.

Advanced surface irrigation techniques such as precise land levelling, gated pipes or cablegation have also been developed to improve the typical poorly distributed application of water under conventional methods by furrow, border or basin irrigation. Experience shows that these techniques are too sophisticated to be adopted by most smallholders in developing countries. Adaptive research in the Middle East and North Africa has shown the difficulties in introducing surface water conservation techniques. Many attempts to introduce gated pipes in Morocco for example have failed for a number of reasons, mainly the poor maintenance of the quaternary earth canals. Considerable efforts for training of farmers in water conservation in surface systems are needed.

Farmers have adopted pressurised systems much faster than surface application techniques. For example, about 65 percent of the farmers in the Jordan valley have shifted from surface to drip irrigation over a 10-year period. Farmers built on-farm storage reservoirs to provide the flexibility required for drip irrigation.

Although there are no statistical data available on it, it is likely that most of the areas irrigated through water conservation techniques are served from wells or piped systems. There is still a huge potential for development of water conservation techniques in the areas presently irrigated from groundwater. Application of these techniques will continue to expand with the market for high value crops. However, the adoption of these techniques is limited to the cultivation of certain crops because of technical and financial considerations.

6.7 WATER RECYCLING PROCESS

Recycled water, is former wastewater (sewage) that is treated to remove solids and certain impurities, and used in sustainable landscaping irrigation or to recharge groundwater aquifers. The purpose of these processes is sustainability and water conservation, rather than discharging the treated water to surface waters such as rivers and oceans.

The definition of reclaimed water, as defined by Levine and Asano, is "The end product of wastewater reclamation that meets water quality requirements for biodegradable materials, suspended matter and pathogens." In more recent conventional use, the term refers to water that is not treated as highly in order to offer a way to conserve drinking water. This water is thusly given to uses such as agriculture and sundry industry uses.

Cycled repeatedly through the planetary hydrosphere, all water on Earth is recycled water. But, typically when we hear the term "recycled water" or "reclaimed water" it means wastewater that is sent from our home or business through a pipeline system to a treatment facility where is treated to a level consistent with its intended use. It is then routed directly to a recycled water system for uses such as irrigation or industrial cooling.

The recycling and recharging is often done by using the treated wastewater for designated municipal sustainable gardening irrigation applications. In most locations, it is intended to only be used for nonpotable uses, such as irrigation, dust control, and fire suppression.

There are examples of communities that have safely used recycled water for many years. Los Angeles County's sanitation districts have provided treated wastewater for landscape irrigation in parks and golf courses since 1929. The first reclaimed water facility inCalifornia was built at San Francisco's Golden Gate Park in 1932. The Irvine Ranch Water District (IRWD) was the first water district in California to receive an unrestricted use permit from the state for its recycled water; such a permit means that water can be used for any purpose except drinking. IRWD maintains one of the largest recycled water systems in the nation with more than 400 miles serving more than 4,500 metered connections. The Irvine Ranch Water District and Orange County Water District in Southern California are established leaders in recycled water. Further, the Orange County Water District, located in in Orange County, and in other locations throughout the world such as Singapore, water is given more advanced treatments and is used indirectly for drinking.

In spite of quite simple methods that incorporate the principles of water-sensitive urban design (WSUD) for easy recovery of stormwater runoff, there remains a common perception that reclaimed water must involve sophisticated and technically complex treatment systems, attempting to recover the most complex and degraded types of sewage. As this effort is supposedly driven bysustainability factors, this type of implementation should inherently be associated with point source solutions, where it is most economical to achieve the expected outcomes. Harvesting of stormwater or rainwater can be an extremely simple to comparatively complex, as well as energy and chemical intensive, recovery of more contaminated sewage.

Storm and sanitary sewers were necessarily developed along with the growth of cities. By the 1840s the luxury of indoor plumbing, which mixes human waste with water and flushes it away, eliminated the need for cesspools. Odor was considered the big problem in waste disposal and to address it, sewage could be drained to a lagoon, or "settled" and the solids removed, to be disposed of separately. This process is now called "primary treatment" and the settled solids are called "sludge."

At the end of the 19th century, since primary treatment still left odor problems, it was discovered that bad odors could be prevented by introducing oxygen into the decomposing sewage. This was the beginning of the biological aerobic and anaerobic treatments which are fundamental to waste water processes. By the 1920s, it became necessary to further control the pollution caused by the large quantities of human and industrial liquid wastes which were being piped into rivers and oceans, and modern treatment plants were being built in the US and other industrialized nations by the 1930s.

Designed to make water safe for fishing and recreation, the Clean Water Act of 1972 mandated elimination of the discharge of untreated waste from municipal and industrial sources, and the US federal government provided billions of dollars in grants for building sewage treatment plants around the country. Modern treatment plants, usually using oxidation and/or chlorination in addition to primary and secondary treatment, were required to meet certain standards.

Current treatment improves the quality of separated wastewater solids or sludge. The separated water is given further treatment considered adequate for non potable use by local agencies, and discharged into bodies of water, or reused as reclaimed water. In places like Florida, where it is necessary to avoid nutrient overload of sensitive receiving water, reuse of treated or reclaimed water can be more economically feasible than meeting the higher standards for surface water disposal mandated by the Clean Water Act

6.7.1 Maximum Water Recovery

To determine maximum water recovery there are various techniques that have been developed by researchers; for maximum water reuse/reclamation/recovery strategies such as water pinch analysis. The techniques help a user to target the minimum freshwater consumption and wastewater target. It also helps in designing the network that achieves the target. This provides a benchmark to be used by users in improving their water systems.

6.7.2 Benefits

The cost of reclaimed water exceeds that of potable water in many regions of the world, where a fresh water supply is plentiful. However, reclaimed water is usually sold to citizens at a cheaper rate to encourage its use. As fresh water supplies become limited from distribution costs, increased population demands, or climate change reducing sources, the cost ratios will evolve also.

Using reclaimed water for non-potable uses saves potable water for drinking, since less potable water will be used for non-potable uses.

It sometimes contains higher levels of nutrients such as nitrogen, phosphorus and oxygen which may somewhat help fertilize garden and agricultural plants when used for irrigation.

The usage of water reclamation decreases the pollution sent to sensitive environments. It can also enhance wetlands, which benefits the wildlife depending on that eco-system. For instance, The San Jose/Santa Clara Water Pollution Control Plant instituted a water recycling program to protect the San Francisco Bay area's natural salt water marshes.

6.7.3 Concerns

Reclaimed water is highly engineered for safety and reliability so that the quality of reclaimed water is more predictable than many existing surface and groundwater sources. Reclaimed water is considered safe when appropriately used. Reclaimed water planned for use in recharging aquifers or augmenting surface water receives adequate and reliable treatment before mixing with naturally occurring water and undergoing natural restoration processes. Some of this water eventually becomes part of drinking water supplies.

A water quality study published in 2009 compared the water quality differences of reclaimed/recycled water, surface water, and groundwater. Results indicate that reclaimed water, surface water, and groundwater are more similar than dissimilar with regard to constituents. The researchers tested for 244 representative constituents typically found in water. When detected, most constituents were in the parts per billion and parts per trillion range. DEET (a bug repellant), and Caffeine were found in all water types and virtually in all samples. Triclosan (in anti-bacterial soap & toothpaste) was found in all water types, but detected in higher levels (parts per trillion) in reclaimed water than in surface or groundwater. Very few hormones/steroids were detected in samples, and when detected were at very low levels. Haloacetic acids (a disinfection by-product) were found in all types of samples, even groundwater. The largest difference between reclaimed water and the other waters appears to be that reclaimed water has been disinfected and thus has disinfection by-products (due to chlorine use).

A 2005 study titled "Irrigation of Parks, Playgrounds, and Schoolyards with Reclaimed Water" found that there had been no incidences of illness or disease from either microbial pathogens or chemicals, and the risks of using using reclaimed water for irrigation are not measurably different from irrigation using potable water. Studies by the National Academies of Science, the Monterey Regional Water Pollution Control Agency, and others have found reclaimed water to be safe for agricultural use.

There is debate about possible health and environmental effects. To address these concerns, A Risk Assessment Study of potential health risks of recycled water and comparisons to conventional Pharmaceuticals and Personal Care Product (PPCP) exposures was conducted by the WateReuse Research Foundation. For each of four scenarios in which people come into contact with recycled water used for irrigation - children on the playground, golfers, and landscape, and agricultural workers - the findings from the study indicate that it could take anywhere from a few years to millions of years of exposure to nonpotable recycled water to reach the same exposure to PPCPs that we get in a single day through routine activities.

6.8 DISTRIBUTION AND DEMAND

Reclaimed water is often distributed with a dual piping network that keeps reclaimed water pipes completely separate from potable water pipes. In the United States, reclaimed water is always distributed in lavender (light purple) pipes to distinguish it from potable water. The use of the color purple for pipes carrying recycled water was pioneered by the Irvine Ranch Water District in Irvine, California.

In many cities using reclaimed water, it is now in such demand that consumers are only allowed to use it on assigned days. Some cities that previously offered unlimited reclaimed water at a flat rate are now beginning to charge citizens by the amount they use.

6.9 TESTING STANDARDS

Reclaimed water is not regulated by the Environmental Protection Agency (EPA) but by the states, using standards formulated decades ago. Newer information shows serious public health concerns about pathogens in the water. Many pathogens cannot be detected by currently used tests.

Recent literature also questions the validity of testing for "indicator organisms" instead of pathogens. Nor do present standards consider interactions of heavy metals and pharmaceuticals which may foster the development of drug resistant pathogens in waters derived from sewage.

6.9.1 Potable Uses

In most locations, reclaimed water is not directly mixed with potable (drinking) water for several reasons:

- Utilities providing reclaimed water for nonpotable uses do not treat the water to drinking water standards.

- Varying amounts of pathogens, pharmaceutical chemicals (e.g., hormones from female hormonal contraception) and other trace chemicals are able to pass through the treatment and filtering process, potentially causing danger to humans. Modern technologies such as reverse osmosis may help to somewhat overcome this problem. An experiment by the University of New South Wales reportedly showed a reverse osmosis system removed ethinylestradiol and paracetamol from the wastewater, even at 1000 times the expected concentration.
- Drinking water standards were developed for natural ground water, and are not appropriate for identifying contaminants in reclaimed water. In addition to pathogens, and organic and endocrine disrupting chemicals, a large number of compounds may be present in reclaimed water. They cannot all be tested for, and there is a paucity of toxicity information on many of the compounds.

Because of this, state regulatory agencies do not allow reclaimed water to be used for drinking, bathing, or filling swimming pools. They also warn those who use reclaimed water for irrigation to place a sign on their property warning people not to drink from the irrigation system, and to not use it directly on fruits or vegetables.

Aboard the International Space Station, astronauts have been able to drink recycled urine due to the introduction of the ECLSS system. The system cost $250 million and has been working since May 2009. The system recycles wastewater and urine back into potable water used for drinking, food preparation, and oxygen generation. This cuts back on the need for resupplying the space station so often.

6.9.2 Indirect Potable Reuse

Some municipalities are now investigating Indirect Potable Use (IPU) of reclaimed water. For example, reclaimed water may be pumped into (subsurface recharge) or percolated down to (surface recharge) groundwater aquifers, pumped out, treated again, and finally used as drinking water. This technique may also be referred to as groundwater recharging.

6.9.3 Unplanned Indirect Potable Reuse

Unplanned Indirect Potable Use has existed even before the introduction of reclaimed water. Many cities already use water from rivers that contain effluent discharged from upstream sewage treatment plants. There are many large towns on the River Thamesupstream of London (Oxford, Reading, Swindon, Bracknell) that discharge their treated sewage into the river, which is used to supply London with water downstream.

This phenomenon is also observed in the United States, where the Mississippi River serves as both the destination of sewage treatment plant effluent and the source of potable water.

Research conducted in the 1960s by the London Metropolitan Water Board demonstrated that the maximum extent of recycling water is about 11 times before the taste of water induces nausea in sensitive individuals. This is caused by the build up of inorganic ions such as Cl, SO_4, K and Na, which are not removed by conventional sewage treatment.

6.10 SPACE TRAVEL

Wastewater reclamation can be especially important in relation to human spaceflight.

- In 1998, NASA announced it had built a human waste reclamation bioreactor designed for use in the International Space Station and a manned Mars mission. Human urine and feces are input into one end of the reactor and pure oxygen, pure water, and compost(humanure) are output from the other end. The soil could be used for growing vegetables, and the bioreactor also produceselectricity.

6.11 TREATMENT IMPROVEMENTS

As world populations require both more clean water and better ways to dispose of wastewater, the demand for water reclamation will increase. Future success in water reuse will depend on whether this can be done without adverse effects on human health and the environment.

In the United States, reclaimed waste water is generally treated to secondary level when used for irrigation, but there are questions about the adequacy of that treatment. Some leading scientists in the main water society, AWWA, have long believed that secondary treatment is insufficient to protect people against pathogens, and recommend adding at least membrane filtration, reverse osmosis, ozonation, or other advanced treatments for irrigation water.

Recent Advances in Reverse Osmosis have been in different countries, but have consistently produced very high quality water all the same. In Singapore, reclaimed water, also known as NEWater has become cleaner than the government issue tap water. Also, according to Bartels, the Bedok Demonstration Plant, which uses RO membranes, has successfully run for the past 3 years, producing high quality wastewater all the while.

Seepage of nitrogen and phosphorus into ground and surface water is also becoming a serious problem, and will probably lead to at least tertiary treatment of reclaimed to remove nutrients in future. Even using secondary treatment, water quality can be improved. Testing for pathogens using Polymerase Chain Reaction (PCR) instead of older culturing techniques, and changing the discredited fecal coloform "indicator organism" standard would be improvements.

In a large study treatment plants showed that they could significantly reduce the numbers of parasites in effluent, just by making adjustments to the currently used process. But, even using the best of current technology, risk of spreading drug resistance in the environment through wastewater effluent, would remain.

Some scientists have suggested that there need to be basic changes in treatment, such as using bacteria to degrade waste based on nitrogen (urine) and not just carbonaceous (fecal) waste, saying that this would greatly improve effectiveness of treatment. Currently designed plants do not deal well with contaminants in solution (e.g. pharmaceuticals). "Dewatering" solids is a major problem. Some wastes could be disposed of without mixing them with water to begin with. In an interesting innovation, solids (sludge) could be removed before entering digesters and burned into a gas that could be used to run engines.

Emerging disinfection technologies include ultrasound, pulse arc electrohydrolic discharge, and bank filtration. Another issue is concern about weakened mandates for pretreatment of industrial wastes before they are made part of the municipal waste stream.Some also believe that hospitals should treat their own wastes. The safety of drinking reclaimed water which has been given advanced treatment and blended with other waters, remains controversial.

6.11.1 Other Alternatives

In urban areas where climate change has threatened long-term water security and reduced rainfall over catchment areas, using reclaimed water for indirect potable use may be superior to other water supply augmentation methods. One other commonly used option is seawater desalination. Recycling wastewater and desalinating seawater may have many of the same disadvantages, including high costs of water treatment, infrastructure construction, transportation, and waste disposal problems. Although the best option varies from region to region, desalination is often superior economically, as reclaimed water usually requires a dual piping network, often with additional storage tanks, when used for nonpotable use.

A less elaborate alternative to reclaimed water is a greywater system. Greywater is wastewater that has been used in sinks, baths, showers, or washing machines, but does not contain sewage (see blackwater) and has not been treated at the same levels as recycled water. In a home system, treated or untreated greywater may be used to flush toilets or for irrigation. Some systems now exist which directly use greywater from a sink to flush a toilet or even combine the two into one piece of furniture.

Perhaps the simplest option is a rainwater harvesting system. Although there are concerns about the quality of rainwater in urban areas, due to air pollution and acid rain, many systems exist now to use untreated rainwater for nonpotable uses or treated rainwater for direct potable use. Urban design systems which incorporate rainwater harvesting and reduce runoff are known as Water Sensitive Urban Design (WSUD) in Australia, Low Impact Development (LID) in United states and Sustainable Urban Drainage Systems (SUDS) in United Kingdom. There are also concerns about rainwater harvesting systems reducing the amount of run-off entering natural bodies of water.

6.12 WORLDWIDE APPLICATIONS AND ACCEPTANCE

As of 2010, Israel leads the world in the proportion of water it recycles. Israel treats 80% of its sewage (400 billion liters a year), and 100% of the sewage from the Tel Aviv metropolitan area is treated and reused as irrigation water for agriculture and public works. The remaining sludge is currently pumped into the Mediterranean, however a new bill has passed stating a conversion to treating the sludge to be used as manure. Only 20% of the treated water is lost (due to evaporation, leaks, overflows and seeping). The recycled water allows farmers to plan ahead and not be limited by water shortages. There are many levels of treatment, and many different ways of treating the water-which leads to a big difference in the quality of the end product. The best quality of reclaimed sewage water comes from adding a gravitational filtering step, after the chemical and biological cleansing. This method uses small ponds in which the water seeps through the sand into the aquifer in about 400 days, then is pumped out as clear purified water. This is nearly the same process used in the space station water recycling system, which turns urine and feces into purified drinking water, oxygen and manure. To add to the efficiency of the Israeli system - the reclaimed sewage water may be mixed with reclaimed sea water (Plans are in action to increase the desalinization program up to 50% of the countries usage by 2013 - 600 billion liters of drinkable sea water a year), along with aquifer water and fresh sweet lake water - monitored by computer to account for the nationwide needs and input. This action reduced the outdated risk of salt and mineral percentages in the water. Plans to implement this overall usage of reclaimed water for drinking are discouraged by the psychological preconception of the public for the quality of reclaimed water, and the fear of its origin. As of today, all the reclaimed sewage water in Israel is used for agricultural and land improvement purposes.

The second largest waste reclamation program in the world is in Spain, where 12% of the nation's waste is treated.

The leaders in use of reclaimed water in the U.S. are Florida and California, with Irvine Ranch Water District as one of the leading developers. They were the first district to approve the use of reclaimed water for in-building piping and use in flushing toilets.

In January 2012 U.S. National Research Council report, a committee of independent experts found that expanding the reuse of municipal wastewater for irrigation, industrial uses, and drinking water augmentation could significantly increase the United States' total available water resources. The committee noted that a portfolio of treatment options is available to mitigate water quality issues in reclaimed water. The report also includes a risk analysis that suggests the risk of exposure to certain microbial and chemical contaminants from drinking reclaimed water is not any higher than the risk from drinking water from current water treatment systems-and in some cases, may be orders of magnitude lower.

The report concludes that adjustments to the federal regulatory framework could enhance public health protection and increase public confidence in water reuse.

As Australia continues to battle the 7-10-year drought, nationwide, reclaimed effluent is becoming a popular option. Two major capital cities in Australia, Adelaide and Brisbane, have already committed to adding reclaimed effluent to their dwindling dams. Brisbane has been seen as a leader in this trend, and other cities and towns will review the Western Corridor Recycled Water Project once completed. Goulbourn, Canberra, Newcastle, and Regional Victoria, Australia are already considering building a reclaimed effluent process.

According to an EU-funded study, "Europe and the Mediterranean countries are lagging behind" California, Japan, and Australia "in the extent to which reuse is being taken up." According to the study "the concept (of reuse) is difficult for the regulators and wider public to understand and accept."

6.13 WATER EFFICIENCY

Water efficiency can be defined as:

1. The accomplishment of a function, task, process, or result with the minimal amount of water feasible;
2. An indicator of the relationship between the amount of water required for a particular purpose and the amount of water used or delivered.

Though the two are often used interchangeably, there is a difference between water conservation and water efficiency. Water efficiency differs from water conservation in that it focuses on reducing waste. A proposition is that the key for efficiency is reducing waste, not restricting use. It also emphasises the influence consumers can have in water efficiency by making small behavioural changes to reduce water wastage and by choosing more water efficient products. Examples of water efficient steps includes fixing leaking taps, taking showers rather than baths, installing displacements devices inside toilet cisterns, and using dishwashers and washing machines with full loads. These are things that fall under the definition of water efficiency, as their purpose is to obtain the desired result or level of service with the least necessary water.

According to the Second UN World Water Development Report, if present levels of consumption continue, two-thirds of the global population will live in areas of water stress by 2025. Increasing human demand for water coupled with the effects of climate change mean that the future of our water supply is not secure. As of now, 2.6 billion people do not have safe drinking water. Added to this, are the changes in climate, population growth and lifestyles. The changes in human lifestyle and activities require more water per capita. This tightens the competition for water amongst agricultural, industrial, and human consumption.

6.13.1 Residential

According to Waterwise, here are some simple ways to be more water efficient at home:

- Turning off the tap while brushing teeth- a running tap can waste over six liters per minute.
- Putting a "hippo" or other displacement device into the toilet cistern.
- Fixing dripping taps. A dripping tap wastes thousands of litres of water a year.
- Using a full load in the dishwasher and washing machine. A person should be sure to buy a water efficient model when buying a new machine.
- Having a short shower instead of a bath.
- Washing fruits and vegetables in a bowl rather than under a running tap.
- Using the leftover water to feed houseplants.
- Using a watering can or a hosepipe with a trigger nozzle instead of a sprinkler.
- Using a bucket and sponge when washing the car rather than a running hosepipe.
- Using little amount of water at every time
- We should wash our cloths in washing machine rather than washing with hands

6.13.2 Manufacturing

According to Save water!, these are solutions useful to manufacturers:

- **Identifying and eliminating wastage (such as leaks) and inefficient processes (such as continual spray devices on stop-start production lines):** This may be the most low cost area for water savings, as it involves minimal capital outlay. Savings can be made through implementing procedural changes, such as cleaning plant areas with brooms rather than water.
- **Changing processes and plant machinery:** A retrofit of key plant equipment may increase efficiency. Alternatively, upgrades to more efficient models can be factored in to planned maintenance and replacement schedules.
- **Reusing wastewater:** As well as saving on mains water, this option may improve the reliability of supply, whilst reducing trade waste charges and associated environmental risks.

6.13.3 Waterless Products

- Using waterless car wash products to wash cars, boats, motorcycles and bicycles. This could save up to 150 gallons of water per wash.

6.13.4 Utilities

According to US EPA, here are some ideas for communities and utilities:

- Implementing a water-loss management program (e.g. locate and repair leaks).
- Utilities should strive for universal metering.
- Ensuring that fire hydrants are tamper proof.
- Equipment changes - Setting a good example by using water efficient equipment.
- Installing faucet aerators and low flow shower heads in municipal buildings.
- Replace worn out plumbing fixtures, appliances and equipment with water-saving models.
- Minimizing the water used in space cooling equipment in accordance with manufacturer's recommendations. Shut off cooling units when not needed.

Utilities can also modify their billing software to track customers who have taken advantage of various utility sponsored water conservation initiatives (toilet rebates, irrigation rebates, etc.) to see which initiatives provide the greatest water savings for the least cost.

7

TREATMENT PLANTS

7.1 OBJECTIVES

After studying this chapter you will be able to understand:

- Sewage treatment plants.
- Sewage characteristics.
- Types of treatment process.
- Biological nutrient removal.
- Treatment methods
- Brine management

7.2 INTRODUCTION

Wastewater treatment plants may be distinguished by the type of wastewater to be treated. There are numerous processes that can be used to treat wastewater depending on the type and extent of contamination. The treatment steps include physical, chemical and biological treatment processes.

Types of wastewater treatment plants include:

- Sewage treatment plants
- Industrial wastewater treatment plants
- Agricultural wastewater treatment plants
- Leachate treatment plants

7.3 SEWAGE TREATMENT PLANTS

Sewage treatment (or domestic wastewater treatment, municipal wastewater treatment) is a type of wastewater treatment which aims to remove contaminants from sewage to produce an effluent that is suitable for discharge to the surrounding environment or an intended reuse application, thereby preventing water pollution from raw sewage discharges. Sewage contains wastewater from households and businesses and possibly pre-treated industrial wastewater. There are a high number of sewage treatment processes to choose from. These can range from decentralized systems (including on-site treatment systems) to large centralized systems involving a network of pipes and pump stations (called sewerage) which convey the sewage to a treatment plant. For cities that have a combined sewer, the sewers will also carry urban runoff (stormwater) to the sewage treatment plant. Sewage treatment often involves two main stages, called primary and secondary treatment, while advanced treatment also incorporates a tertiary treatment stage with polishing processes and nutrient removal. Secondary treatment can reduce organic matter (measured as biological oxygen demand) from sewage, using aerobic or anaerobic biological processes.

Fig. 1. Aeration tank of an activated sludge process at the wastewater treatment plant in Dresden-Kaditz, Germany

A large number of sewage treatment technologies have been developed, mostly using biological treatment processes. Engineers and decision makers need to take into account technical and economical criteria, as well as quantitative and qualitative aspects of each alternative when choosing a suitable technology. Often, the main criteria for selection are: desired effluent quality, expected construction and operating costs, availability of land, energy requirements and sustainability aspects. In developing countries and in rural areas with low population densities, sewage is often treated by various on-site sanitation systems and not conveyed in sewers. These systems include septic tanks connected to drain fields, on-site sewage systems (OSS), vermifilter systems and many more. On the other hand, advanced and relatively expensive sewage treatment plants in cities that can afford them may include tertiary treatment with disinfection and possibly even a fourth treatment stage to remove micropollutants.

At the global level, an estimated 52% of sewage is treated. However, sewage treatment rates are highly unequal for different countries around the world. For example, while high-income countries treat approximately 74% of their sewage, developing countries treat an average of just 4.2%.

Sewage treatment (or domestic wastewater treatment, municipal wastewater treatment) is a type of wastewater treatment which aims to remove contaminants from sewage to produce an effluent that is suitable for discharge to the surrounding environment or an intended reuse application, thereby preventing water pollution from raw sewage discharges. Sewage contains wastewater from households and businesses and possibly pre-treated industrial wastewater. There are a high number of sewage treatment processes to choose from. These can range from decentralized systems (including on-site treatment systems) to large centralized systems involving a network of pipes and pump stations (called sewerage) which convey the sewage to a treatment plant. For cities that have a combined sewer, the sewers will also carry urban runoff (stormwater) to the sewage treatment plant. Sewage treatment often involves two main stages, called primary and secondary treatment, while advanced treatment also incorporates a tertiary treatment stage with polishing processes and nutrient removal. Secondary treatment can reduce organic matter (measured as biological oxygen demand) from sewage, using aerobic or anaerobic biological processes.

At the global level, an estimated 52% of sewage is treated. However, sewage treatment rates are highly unequal for different countries around the world. For example, while high-income countries treat approximately 74% of their sewage, developing countries treat an average of just 4.2%.

The treatment of sewage is part of the field of sanitation. Sanitation also includes the management of human waste and solid waste as well as stormwater (drainage) management. The term "sewage treatment plant" is often used interchangeably with the term "wastewater treatment plant".

7.3.1 Purposes and Overview

The overall aim of treating sewage is to produce an effluent that can be discharged to the environment while causing as little water pollution as possible, or to produce an effluent that can be reused in a useful manner. This is achieved by removing contaminants from the sewage. It is a form of waste management.

With regards to biological treatment of sewage, the treatment objectives can include various degrees of the following: transform dissolved and particulate biodegradable components (especially organic matter) into acceptable end products, transform and remove nutrients (nitrogen and phosphorus), remove or inactivate pathogenic organisms, and remove specific trace organic constituents (micropollutants).

Some types of sewage treatment produce sewage sludge which can be treated before safe disposal or reuse. Under certain circumstances, the treated sewage sludge might be termed "biosolids" and can be used as a fertilizer.

7.3.2 Sewage Characteristics

Typical values for physical-chemical characteristics of raw sewage in developing countries have been published as follows: 180 g/person/d for total solids (or 1100 mg/L when expressed as a concentration), 50 g/person/d for BOD (300 mg/L), 100 g/person/d for COD (600 mg/L), 8 g/person/d for total nitrogen (45 mg/L), 4.5 g/person/d for ammonia-N (25 mg/L) and 1.0 g/person/d for total phosphorus (7 mg/L). The typical ranges for these values are: 120–220 g/person/d for total solids (or 700–1350 mg/L when expressed as a concentration), 40–60 g/person/d for BOD (250–400 mg/L), 80–120 g/person/d for COD (450–800 mg/L), 6–10 g/person/d for total nitrogen (35-60 mg/L), 3.5–6 g/person/d for ammonia-N (20–35 mg/L) and 0.7–2.5 g/person/d for total phosphorus (4–15 mg/L).

For high income countries, the "per person organic matter load" has been found to be approximately 60 gram of BOD per person per day. This is called the Population Equivalent (PE) and is also used as a comparison parameter to express the strength of industrial wastewater compared to sewage.

7.3.3 Collection

Sewerage (or sewage system) is the infrastructure that conveys sewage or surface runoff (stormwater, meltwater, rainwater) using sewers. It encompasses components such as receiving drains, manholes, pumping stations, storm overflows, and screening chambers of the combined sewer or sanitary sewer. Sewerage ends at the entry to a sewage treatment plant or at the point of discharge into the environment. It is the system of pipes, chambers, manholes, etc. that conveys the sewage or storm water.

In many cities, sewage (or municipal wastewater) is carried together with stormwater, in a combined sewer system, to a sewage treatment plant. In some urban areas, sewage is carried separately in sanitary sewers and runoff from streets is carried in storm drains. Access to these systems, for maintenance purposes, is typically through a manhole. During high precipitation periods a sewer system may experience a combined sewer overflow event or a sanitary sewer overflow event, which forces untreated sewage to flow directly to receiving waters. This can pose a serious threat to public health and the surrounding environment.

The system of sewers is called sewerage or sewerage system in British English and sewage system in American English.

7.4 TYPES OF TREATMENT PROCESSES

Sewage can be treated close to where the sewage is created, which may be called a "decentralized" system or even an "on-site" system (on-site sewage facility, septic tanks, etc.). Alternatively, sewage can be collected and transported by a network of pipes and pump stations to a municipal treatment plant. This is called a "centralized" system (see also sewerage and pipes and infrastructure).

A large number of sewage treatment technologies have been developed, mostly using biological treatment processes. Very broadly, they can be grouped into high tech (high cost) versus low tech (low cost) options, although some technologies might fall into either category. Other grouping classifications are "intensive" or "mechanized" systems (more compact, and frequently employing high tech options) versus "extensive" or "natural" or "nature-based" systems (usually using natural treatment processes and occupying larger areas) systems. This classification may be sometimes oversimplified, because a treatment plant may involve a combination of processes, and the interpretation of the concepts of high tech and low tech, intensive and extensive, mechanized and natural processes may vary from place to place.

7.4.1 Low Tech, Extensive or Nature-Based Processes

Examples for more low-tech, "natural", often less expensive sewage treatment systems are shown below. They often use little or no energy. Some of these systems do not provide a high level of treatment, or only treat part of the sewage (for example only the toilet wastewater), or they only provide pre-treatment, like septic tanks. On the other hand, some systems are capable of providing a good performance, satisfactory for several applications. Many of these systems are based on natural treatment processes, requiring large areas, while others are more compact. In most cases, they are used in rural areas or in small to medium-sized communities. For example, waste stabilization ponds are a low cost treatment option with practically no energy requirements but they require a lot of land. Due to their technical simplicity, most of the savings (compared with high tech systems) are in terms of operation and maintenance costs.

- Anaerobic digester types and anaerobic digestion, for example:
    - Upflow anaerobic sludge blanket reactor
    - Septic tank
    - Imhoff tank
- Constructed wetland (see also biofilters)
- Decentralized wastewater system
- Nature-based solutions
- On-site sewage facility
- Sand filter
- Vermifilter
- Waste stabilization pond with sub-types:
    - Facultative ponds
    - Anaerobic pond – facultative ponds systems
    - Facultative aerated lagoons

  - o Complete-mix aerated lagoon sedimentation pond systems
  - o High rate ponds
  - o Maturation ponds

Examples for systems that can provide full or partial treatment for toilet wastewater only:

- Composting toilet
- Urine-diverting dry toilet
- Vermifilter toilet

7.4.2 High Tech, Intensive or Mechanized Processes

Examples for more high-tech, intensive or "mechanized", often relatively expensive sewage treatment systems are listed below. Some of them are energy intensive as well. Many of them provide a very high level of treatment. For example, broadly speaking, the activated sludge process achieves a high effluent quality but is relatively expensive and energy intensive.

- Activated sludge systems
- Aerobic granulation
- Aerobic treatment system
- Enhanced biological phosphorus removal
- Expanded granular sludge bed digestion
- Extended aeration
- Filtration
- Membrane bioreactor
- Moving bed biofilm reactor
- Reverse osmosis
- Rotating biological contactor
- Sequencing batch reactor
- Trickling filter
- Ultrafiltration
- Ultraviolet disinfection

7.4.3 Disposal or Treatment Options

There are other process options which may be classified as disposal options, although they can also be understood as basic treatment options. These include: Application of sludge, irrigation, soak pit, leach field, fish pond, floating plant pond, water disposal/ groundwater recharge, surface disposal and storage.

Application of sewage to land can be considered as a form of final disposal or of treatment, or both. It leads to groundwater recharge and/or to evapotranspiration. Land application include slow-rate systems, rapid infiltration, subsurface infiltration, overland flow. It is done by flooding, furrows, sprinkler and dripping. It is a treatment/disposal system that requires a large amount of land per person.

7.5 DESIGN ASPECTS

7.5.1 Population Equivalent

The "per person organic matter load" is a parameter used in the design of sewage treatment plants. This concept is known as population equivalent (PE). The base value used for PE can vary from one country to another. Commonly used definitions used worldwide are: 1 PE equates to 60 gram of BOD per person per day, and it also equals 200 liters of sewage per day. This concept is also used as a comparison parameter to express the strength of industrial wastewater compared to sewage.

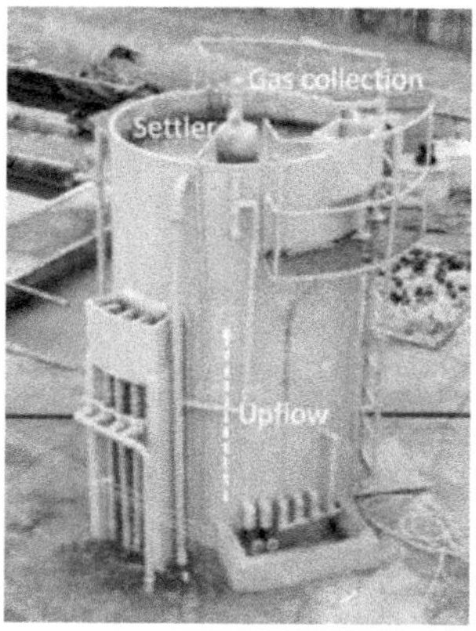

Fig. 2. Upflow anaerobic sludge blanket (UASB) reactor in Brazil (picture from a small-sized treatment plant), Center for Research and Training in Sanitation, Belo Horizonte, Brazil.

7.6 PROCESS SELECTION

When choosing a suitable sewage treatment process, decision makers need to take into account technical and economical criteria, as well as quantitative and qualitative aspects of each alternative. Therefore, each analysis is site-specific. A life cycle assessment (LCA) can be used, and criteria or weightings can be attributed to the various aspects. The final decision may have a degree of subjectivity. A range of publications exist to help with technology selection.

In industrialized countries, the critical items in process selection are, in decreasing order of importance: efficiency, reliability, sludge disposal aspects and land requirements.

In developing countries, the main critical items might be different and revolve more around construction costs, sustainability, simplicity and operational costs.

Choosing the most suitable treatment process is complicated and requires expert inputs, often in the form of feasibility studies. This is because the main important factors to be considered when evaluating and selecting sewage treatment processes are numerous: process applicability, applicable flow, acceptable flow variation, influent characteristics, inhibiting or refractory compounds, climatic aspects, process kinetics and reactor hydraulics, performance, treatment residuals, sludge processing, environmental constraints, chemical product requirements, energy requirements, requirements of other resources, personnel requirements, operating and maintenance requirements, ancillary processes, reliability, complexity, compatibility, area availability.

With regards to environmental impacts the following aspects are included in the selection process: Odors, vector attraction, sludge transportation, sanitary risks, air contamination, soil and subsoil contamination, surface water pollution or groundwater contamination, devaluation of nearby areas, inconvenience to the nearby population.

7.7 ODOR CONTROL

Odors emitted by sewage treatment are typically an indication of an anaerobic or "septic" condition. Early stages of processing will tend to produce foul-smelling gases, with hydrogen sulfide being most common in generating complaints. Large process plants in urban areas will often treat the odors with carbon reactors, a contact media with bio-slimes, small doses of chlorine, or circulating fluids to biologically capture and metabolize the noxious gases. Other methods of odor control exist, including addition of iron salts, hydrogen peroxide, calcium nitrate, etc. to manage hydrogen sulfide levels.

7.7.1 Energy Requirements

The energy requirements vary with type of treatment process as well as sewage strength. For example, constructed wetlands and stabilization ponds have low energy requirements, associated mainly with the occasional presence of pumps and other equipment. On the other hand, the activated sludge process includes an aeration step, which is highly energy consuming. Sewage treatment plants that produce biogas in their sewage sludge treatment process with anaerobic digestion can produce enough energy to meet most of the energy needs of the sewage treatment plant itself. For activated sludge treatment plants in the United States, around 30 percent of the annual operating costs is usually required for energy. Most of this electricity is used for aeration, pumping systems and equipment for the dewatering and drying of sewage sludge. Advanced sewage treatment plants, e.g. for nutrient removal, require more energy than plants that only achieve primary or secondary treatment.

Small rural plants using trickling filters may operate with no net energy requirements, the whole process being driven by gravitational flow, including tipping bucket flow distribution and the desludging of settlement tanks to drying beds. This is usually only practical in hilly terrain and in areas where the treatment plant is relatively remote from housing because of the difficulty in managing odors.

7.8 CO-TREATMENT OF INDUSTRIAL EFFLUENT

In highly regulated developed countries, industrial wastewater usually receives at least pretreatment if not full treatment at the factories themselves to reduce the pollutant load, before discharge to the sewer. The pretreatment has the following aims: to remove constituents that may pose risks to the sewerage system and its workers; prevent toxic or inhibitory compounds to the microorganisms in the biological stage in the municipal treatment plant; hinder beneficial use of the produced sewage sludge; or that will still be present in the final effluent from the treatment plant. Some industrial wastewater may contain pollutants which cannot be removed by sewage treatment plants. Also, variable flow of industrial waste associated with production cycles may upset the population dynamics of biological treatment units.

7.9 DESIGN ASPECTS OF SECONDARY TREATMENT PROCESSES

7.9.1 Non-Sewered Areas

Urban residents in many parts of the world rely on on-site sanitation systems without sewers, such as septic tanks and pit latrines, and fecal sludge management in these cities is an enormous challenge.

For sewage treatment the use of septic tanks and other On-Site Sewage Facilities (OSSF) is widespread in some rural areas, for example serving up to 20 percent of the homes in the U.S.

7.10 AVAILABLE PROCESS STEPS

Sewage treatment often involves two main stages, called primary and secondary treatment, while advanced treatment also incorporates a tertiary treatment stage with polishing processes. Different types of sewage treatment may utilize some or all of the process steps listed below.

7.10.1 Preliminary Treatment

Preliminary treatment (sometimes called pretreatment) removes coarse materials that can be easily collected from the raw sewage before they damage or clog the pumps and sewage lines of primary treatment clarifiers.

7.10.2 Screening

The influent in sewage water passes through a bar screen to remove all large objects like cans, rags, sticks, plastic packets, etc. carried in the sewage stream. This is most commonly done with an automated mechanically raked bar screen in modern plants serving large populations, while in smaller or less modern plants, a manually cleaned screen may be used.

Fig. 3. Preliminary treatment arrangement at small and medium-sized sewage treatment plants: Manually-cleaned screens and grit chamber (Jales Treatment Plant, São Paulo, Brazil).

The raking action of a mechanical bar screen is typically paced according to the accumulation on the bar screens and/or flow rate. The solids are collected and later disposed in a landfill, or incinerated. Bar screens or mesh screens of varying sizes may be used to optimize solids removal. If gross solids are not removed, they become entrained in pipes and moving parts of the treatment plant, and can cause substantial damage and inefficiency in the process.

7.10.3 Grit Removal

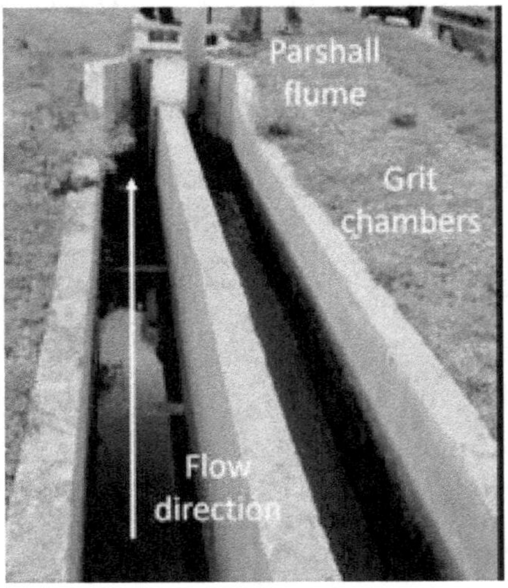

Fig. 4. Preliminary treatment: Horizontal flow grit chambers at a sewage treatment plant in Juiz de Fora, Minas Gerais, Brazil.

Grit consists of sand, gravel, rocks, and other heavy materials. Preliminary treatment may include a sand or grit removal channel or chamber, where the velocity of the incoming sewage is reduced to allow the settlement of grit. Grit removal is necessary to (1) reduce formation of deposits in primary sedimentation tanks, aeration tanks, anaerobic digesters, pipes, channels, etc. (2) reduce the frequency of tank cleaning caused by excessive accumulation of grit; and (3) protect moving mechanical equipment from abrasion and accompanying abnormal wear. The removal of grit is essential for equipment with closely machined metal surfaces such as comminutors, fine screens, centrifuges, heat exchangers, and high pressure diaphragm pumps.

Grit chambers come in three types: horizontal grit chambers, aerated grit chambers, and vortex grit chambers. Vortex grit chambers include mechanically induced vortex, hydraulically induced vortex, and multi-tray vortex separators. Given that traditionally, grit removal systems have been designed to remove clean inorganic particles that are greater than 0.210 millimetres (0.0083 in), most of the finer grit passes through the grit removal flows under normal conditions. During periods of high flow deposited grit is resuspended and the quantity of grit reaching the treatment plant increases substantially. It is therefore important that the grit removal system not only operates efficiently during normal flow conditions but also under sustained peak flows when the greatest volume of grit reaches the plant.

7.10.4 Flow Equalization

Equalization basins can be used to achieve flow equalization, with the aim to reduce peak dry-weather flows or peak wet-weather flows in the case of combined sewer systems. The benefits are performance improvements of the biological treatment processes, the secondary clarifiers and any effluent filtration equipment. Disadvantages include the basins' capital cost and space requirements. Basins can also provide a place to temporarily hold, dilute and distribute batch discharges of toxic or high-strength wastewater which might otherwise inhibit biological secondary treatment (such was wastewater from portable toilets or fecal sludge that is brought to the sewage treatment plant in vacuum trucks). Flow equalization basins require variable discharge control, typically include provisions for bypass and cleaning, and may also include aerators and odor control.

7.10.5 Fat and Grease Removal

In some larger plants, fat and grease are removed by passing the sewage through a small tank where skimmers collect the fat floating on the surface. Air blowers in the base of the tank may also be used to help recover the fat as a froth. Many plants, however, use primary clarifiers with mechanical surface skimmers for fat and grease removal.

7.11 PRIMARY TREATMENT OF WATER

Primary treatment is the "removal of a portion of the suspended solids and organic matter from the sewage". It consists of allowing sewage to pass slowly through a basin where heavy solids can settle to the bottom while oil, grease and lighter solids float to the surface and are skimmed off. These basis are called "primary sedimentation tanks" or "primary clarifiers" and typically have a hydraulic retention time (HRT) of 1.5 to 2.5 hours. The settled and floating materials are removed and the remaining liquid may be discharged or subjected to secondary treatment. Primary settling tanks are usually equipped with mechanically driven scrapers that continually drive the collected sludge towards a hopper in the base of the tank where it is pumped to sludge treatment facilities.

Fig. 5. Rectangular primary settling tanks at a sewage treatment plant in Oregon, USA.

Sewage treatment plants that are connected to a combined sewer system sometimes have a bypass arrangement after the primary treatment unit. This means that during very heavy rainfall events, the secondary and tertiary treatment systems can be bypassed to protect them from hydraulic overloading, and the mixture of sewage and storm-water receives primary treatment only.

Primary sedimentation tanks can be expected to remove 50-70% of the suspended solids and 25-40% of the biological oxygen demand (BOD).

7.12 SECONDARY TREATMENT OF WATER

Simplified process flow diagram for a typical large-scale treatment plant using the activated sludge process.

The main processes involved in secondary sewage treatment are designed to remove as much of the solid material as possible. They use biological processes to digest and remove the remaining soluble material, especially the organic fraction. This can be done with either suspended-growth or biofilm processes. The microorganisms that feed on the organic matter present in the sewage grow and multiply, constituting the biological solids, or biomass. These grow and group together in the form of flocs or biofilms and, in some specific processes, as granules. In several treatment processes, the biological floc or biofilm and remaining fine solids can then be settled as a sludge, leaving a liquid substantially free of solids, and with a greatly reduced concentration of pollutants.

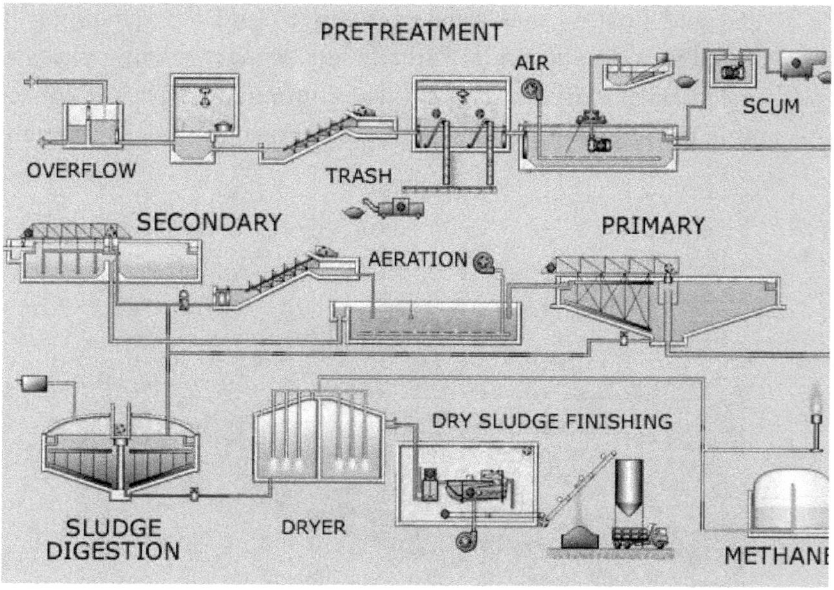

Fig. 6. Secondary treatement.

Secondary treatment can reduce organic matter (measured as biological oxygen demand) from sewage, using aerobic or anaerobic processes. The organisms involved in these processes are sensitive to the presence of toxic materials, although these are not expected to be present at high concentrations in typical municipal sewage.

7.13 TERTIARY TREATMENT OF WATER

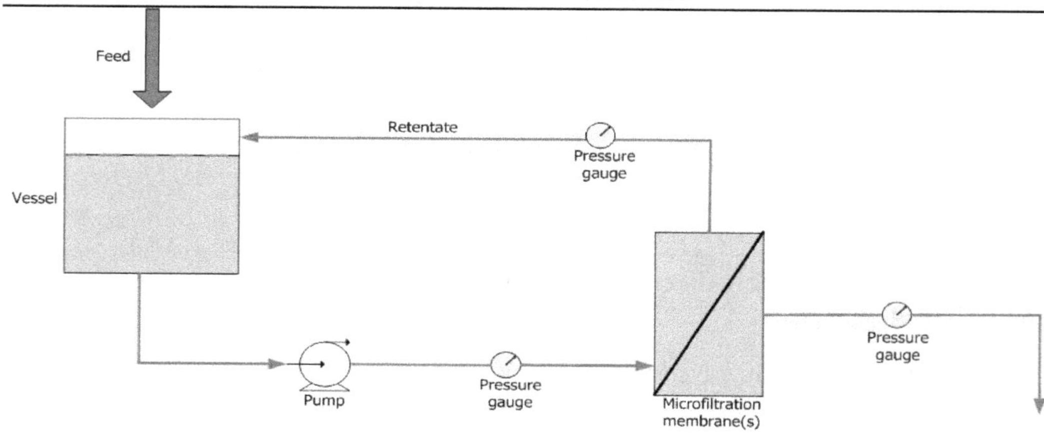

Fig. 7. Overall setup for a micro filtration system

Advanced sewage treatment generally involves three main stages, called primary, secondary and tertiary treatment but may also include intermediate stages and final polishing processes. The purpose of tertiary treatment (also called "advanced treatment") is to provide a final treatment stage to further improve the effluent quality before it is discharged to the receiving water body or reused. More than one tertiary treatment process may be used at any treatment plant. If disinfection is practiced, it is always the final process. It is also called "effluent polishing". Tertiary treatment may include biological nutrient removal (alternatively, this can be classified as secondary treatment), disinfection and removal of micropollutants, such as environmental persistent pharmaceutical pollutants.

Tertiary treatment is sometimes defined as anything more than primary and secondary treatment in order to allow discharge into a highly sensitive or fragile ecosystem such as estuaries, low-flow rivers or coral reefs. Treated water is sometimes disinfected chemically or physically (for example, by lagoons and microfiltration) prior to discharge into a stream, river, bay, lagoon or wetland, or it can be used for the irrigation of a golf course, greenway or park. If it is sufficiently clean, it can also be used for groundwater recharge or agricultural purposes.

Sand filtration removes much of the residual suspended matter. Filtration over activated carbon, also called carbon adsorption, removes residual toxins. Micro filtration or synthetic membranes are also used, for example in membrane bioreactors which also remove pathogens.

Settlement and further biological improvement of treated sewage may be achieved through storage in large human-made ponds or lagoons. These lagoons are highly aerobic, and colonization by native macrophytes, especially reeds, is often encouraged.

7.14 DISINFECTION

Disinfection of treated sewage may be attempted to kill pathogens (disease-causing microorganisms) prior to disposal, and is increasingly effective after more elements of the foregoing treatment sequence have been completed. The purpose of disinfection in the treatment of sewage is to substantially reduce the number of pathogens in the water to be discharged back into the environment or to be reused. The effectiveness of disinfection depends on the quality of the water being treated (e.g. turbidity, pH, etc.), the type of disinfection being used, the disinfectant dosage (concentration and time), and other environmental variables. Water with high turbidity will be treated less successfully, since solid matter can shield organisms, especially from ultraviolet light or if contact times are low. Generally, short contact times, low doses and high flows all militate against effective disinfection. Common methods of disinfection include ozone, chlorine, ultraviolet light, or sodium hypochlorite. Monochloramine, which is used for drinking water, is not used in the treatment of sewage because of its persistence.

Chlorination remains the most common form of treated sewage disinfection in many countries due to its low cost and long-term history of effectiveness. One disadvantage is that chlorination of residual organic material can generate chlorinated-organic compounds that may be carcinogenic or harmful to the environment. Residual chlorine or chloramines may also be capable of chlorinating organic material in the natural aquatic environment. Further, because residual chlorine is toxic to aquatic species, the treated effluent must also be chemically dechlorinated, adding to the complexity and cost of treatment.

Ultraviolet (UV) light can be used instead of chlorine, iodine, or other chemicals. Because no chemicals are used, the treated water has no adverse effect on organisms that later consume it, as may be the case with other methods. UV radiation causes damage to the genetic structure of bacteria, viruses, and other pathogens, making them incapable of reproduction. The key disadvantages of UV disinfection are the need for frequent lamp maintenance and replacement and the need for a highly treated effluent to ensure that the target microorganisms are not shielded from the UV radiation (i.e., any solids present in the treated effluent may protect microorganisms from the UV light). In many countries, UV light is becoming the most common means of disinfection because of the concerns about the impacts of chlorine in chlorinating residual organics in the treated sewage and in chlorinating organics in the receiving water.

As with UV treatment, heat sterilization also does not add chemicals to the water being treated. However, unlike UV, heat can penetrate liquids that are not transparent. Heat disinfection can also penetrate solid materials within wastewater, sterilizing their contents. Thermal effluent decontamination systems provide low resource, low maintenance effluent decontamination once installed.

Ozone (O_3) is generated by passing oxygen (O_2) through a high voltage potential resulting in a third oxygen atom becoming attached and forming O_3. Ozone is very unstable and reactive and oxidizes most organic material it comes in contact with, thereby destroying many pathogenic microorganisms. Ozone is considered to be safer than chlorine because, unlike chlorine which has to be stored on site (highly poisonous in the event of an accidental release), ozone is generated on-site as needed from the oxygen in the ambient air. Ozonation also produces fewer disinfection by-products than chlorination. A disadvantage of ozone disinfection is the high cost of the ozone generation equipment and the requirements for special operators. Ozone sewage treatment requires the use of an ozone generator, which decontaminates the water as ozone bubbles percolate through the tank.

Membranes can also be effective disinfectants, because they act as barriers, avoiding the passage of the microorganisms. As a result, the final effluent may be devoid of pathogenic organisms, depending on the type of membrane used. This principle is applied in membrane bioreactors.

7.15 BIOLOGICAL NUTRIENT REMOVAL

Sewage may contain high levels of the nutrients nitrogen and phosphorus. Typical values for nutrient loads per person and nutrient concentrations in raw sewage in developing countries have been published as follows: 8 g/person/d for total nitrogen (45 mg/L), 4.5 g/person/d for ammonia-N (25 mg/L) and 1.0 g/person/d for total phosphorus (7 mg/L). The typical ranges for these values are: 6-10 g/person/d for total nitrogen (35-60 mg/L), 3.5-6 g/person/d for ammonia-N (20-35 mg/L) and 0.7-2.5 g/person/d for total phosphorus (4-15 mg/L).

Fig. 8. Nitrification process tank at an activated sludge plant in the United States.

Excessive release to the environment can lead to nutrient pollution, which can manifest itself in eutrophication. This process can lead to algal blooms, a rapid growth, and later decay, in the population of algae. In addition to causing deoxygenation, some algal species produce toxins that contaminate drinking water supplies.

Ammonia nitrogen, in the form of free ammonia (NH_3) is toxic to fish. Ammonia nitrogen, when converted to nitrite and further to nitrate in a water body, in the process of nitrification, is associated with the consumption of dissolved oxygen. Nitrite and nitrate may also have public health significance if concentrations are high in drinking water, because of a disease called metahemoglobinemia.

Phosphorus removal is important as phosphorus is a limiting nutrient for algae growth in many fresh water systems. Therefore, an excess of phosphorus can lead to eutrophication. It is also particularly important for water reuse systems where high phosphorus concentrations may lead to fouling of downstream equipment such as reverse osmosis.

A range of treatment processes are available to remove nitrogen and phosphorus. Biological nutrient removal (BNR) is regarded by some as a type of secondary treatment process, and by others as a tertiary (or "advanced") treatment process.

7.15.1 Nitrogen Removal

Nitrogen is removed through the biological oxidation of nitrogen from ammonia to nitrate (nitrification), followed by denitrification, the reduction of nitrate to nitrogen gas. Nitrogen gas is released to the atmosphere and thus removed from the water.

Nitrification itself is a two-step aerobic process, each step facilitated by a different type of bacteria. The oxidation of ammonia (NH_4^+) to nitrite (NO_2^-) is most often facilitated by bacteria such as Nitrosomonas spp. ("nitroso" referring to the formation of a nitroso functional group). Nitrite oxidation to nitrate (NO_3^-), though traditionally believed to be facilitated by Nitrobacter spp. (nitro referring the formation of a nitro functional group), is now known to be facilitated in the environment predominantly by Nitrospira spp.

Denitrification requires anoxic conditions to encourage the appropriate biological communities to form. "Anoxic conditions" refers to a situation where oxygen is absent but nitrate is present. Denitrification is facilitated by a wide diversity of bacteria. The activated sludge process, sand filters, waste stabilization ponds, constructed wetlands and other processes can all be used to reduce nitrogen. Since denitrification is the reduction of nitrate to dinitrogen (molecular nitrogen) gas, an electron donor is needed. This can be, depending on the wastewater, organic matter (from the sewage itself), sulfide, or an added donor like methanol. The sludge in the anoxic tanks (denitrification tanks) must be mixed well (mixture of recirculated mixed liquor, return activated sludge, and raw influent) e.g. by using submersible mixers in order to achieve the desired denitrification.

Over time, different treatment configurations for activated sludge processes have evolved to achieve high levels of nitrogen removal. An initial scheme, the Ludzack–Ettinger Process, placed an anoxic treatment zone before the aeration tank and clarifier, using the return activated sludge (RAS) from the clarifier as a nitrate source. The sewage (either raw or as effluent from primary clarification) serves as the electron source for the facultative bacteria to metabolize carbon, using the inorganic nitrate as a source of oxygen instead of dissolved molecular oxygen. This denitrification scheme was naturally limited to the amount of soluble nitrate present in the RAS. Nitrate reduction was limited because RAS rate is limited by the performance of the clarifier.

The "Modified Ludzak–Ettinger Process" (MLE) is an improvement on the original concept, for it recycles mixed liquor from the discharge end of the aeration tank to the head of the anoxic tank to provide a consistent source of soluble nitrate for the facultative bacteria. In this instance, raw sewage continues to provide the electron source, and sub-surface mixing maintains the bacteria in contact with both electron source and soluble nitrate in the absence of dissolved oxygen.

There are other process configurations, including anoxic tanks before and after the aeration tanks, such as variations of the Bardenpho process.

7.15.2 Phosphorus Removal

Studies of United States sewage in the late 1960s estimated mean per capita contributions of 500 grams (18 oz) in urine and feces, 1,000 grams (35 oz) in synthetic detergents, and lesser variable amounts used as corrosion and scale control chemicals in water supplies. Source control via alternative detergent formulations has subsequently reduced the largest contribution, but naturally the phosphorus content of urine and feces remained unchanged.

Phosphorus can be removed biologically in a process called enhanced biological phosphorus removal. In this process, specific bacteria, called polyphosphate-accumulating organisms (PAOs), are selectively enriched and accumulate large quantities of phosphorus within their cells (up to 20 percent of their mass).

Phosphorus removal can also be achieved by chemical precipitation, usually with salts of iron (e.g. ferric chloride) or aluminum (e.g. alum), or lime. This may lead to a higher sludge production as hydroxides precipitate and the added chemicals can be expensive. Chemical phosphorus removal requires significantly smaller equipment footprint than biological removal, is easier to operate and is often more reliable than biological phosphorus removal. Another method for phosphorus removal is to use granular laterite or zeolite.

Some systems use both biological phosphorus removal and chemical phosphorus removal. The chemical phosphorus removal in those systems may be used as a backup system, for use when the biological phosphorus removal is not removing enough phosphorus, or may be used continuously. In either case, using both biological and chemical phosphorus removal has the advantage of not increasing sludge production as much as chemical phosphorus removal on its own, with the disadvantage of the increased initial cost associated with installing two different systems.

Once removed, phosphorus, in the form of a phosphate-rich sewage sludge, may be sent to landfill or used as fertilizer in admixture with other digested sewage sludges. In the latter case, the treated sewage sludge is also sometimes referred to as biosolids.

7.16 FOURTH TREATMENT STAGE

Micropollutants such as pharmaceuticals, ingredients of household chemicals, chemicals used in small businesses or industries, Environmental Persistent Pharmaceutical Pollutants (EPPP) or pesticides may not be eliminated in the commonly used sewage treatment processes (primary, secondary and tertiary treatment) and therefore lead to water pollution. Although concentrations of those substances and their decomposition products are quite low, there is still a chance of harming aquatic organisms. For pharmaceuticals, the following substances have been identified as "toxicologically relevant": substances with endocrine disrupting effects, genotoxic substances and substances that enhance the development of bacterial resistances. They mainly belong to the group of EPPP.

Techniques for elimination of micropollutants via a fourth treatment stage during sewage treatment are implemented in Germany, Switzerland, Sweden and the Netherlands and tests are ongoing in several other countries. Such process steps mainly consist of activated carbon filters that adsorb the micropollutants. The combination of advanced oxidation with ozone followed by Granular Activated Carbon (GAC) has been suggested as a cost-effective treatment combination for pharmaceutical residues. For a full reduction of microplasts the combination of ultrafiltration followed by GAC has been suggested. Also the use of enzymes such as laccase secreted by fungi is under investigation. Microbial biofuel cells are investigated for their property to treat organic matter in sewage.

To reduce pharmaceuticals in water bodies, "source control" measures are also under investigation, such as innovations in drug development or more responsible handling of drugs. In the US, the National Take Back Initiative is a voluntary program with the general public, encouraging people to return excess or expired drugs, and avoid flushing them to the sewage system.

7.17 TREATMENT OF SLUDGE AND DISPOSAL

Fig. 9. View of a belt filter press at the Blue Plains Advanced Wastewater Treatment Plant, Washington, D.C.

Sewage sludge treatment describes the processes used to manage and dispose of sewage sludge produced during sewage treatment. Sludge treatment is focused on reducing sludge weight and volume to reduce transportation and disposal costs, and on reducing potential health risks of disposal options. Water removal is the primary means of weight and volume reduction, while pathogen destruction is frequently accomplished through heating during thermophilic digestion, composting, or incineration. The choice of a sludge treatment method depends on the volume of sludge generated, and comparison of treatment costs required for available disposal options. Air-drying and composting may be attractive to rural communities, while limited land availability may make aerobic digestion and mechanical dewatering preferable for cities, and economies of scale may encourage energy recovery alternatives in metropolitan areas.

Fig. 10. Mechanical dewatering of sewage sludge with a centrifuge at a large sewage treatment plant (Arrudas Treatment Plant, Belo Horizonte, Brazil)

Sludge is mostly water with some amounts of solid material removed from liquid sewage. Primary sludge includes settleable solids removed during primary treatment in primary clarifiers. Secondary sludge is sludge separated in secondary clarifiers that are used in secondary treatment bioreactors or processes using inorganic oxidizing agents. In intensive sewage treatment processes, the sludge produced needs to be removed from the liquid line on a continuous basis because the volumes of the tanks in the liquid line have insufficient volume to store sludge. This is done in order to keep the treatment processes compact and in balance (production of sludge approximately equal to the removal of sludge). The sludge removed from the liquid line goes to the sludge treatment line. Aerobic processes (such as the activated sludge process) tend to produce more sludge compared with anaerobic processes. On the other hand, in extensive (natural) treatment processes, such as ponds and constructed wetlands, the produced sludge remains accumulated in the treatment units (liquid line) and is only removed after several years of operation.

Sludge treatment options depend on the amount of solids generated and other site-specific conditions. Composting is most often applied to small-scale plants with aerobic digestion for mid-sized operations, and anaerobic digestion for the larger-scale operations. The sludge is sometimes passed through a so-called pre-thickener which de-waters the sludge. Types of pre-thickeners include centrifugal sludge thickeners, rotary drum sludge thickeners and belt filter presses. Dewatered sludge may be incinerated or transported offsite for disposal in a landfill or use as an agricultural soil amendment.

7.18 ENVIRONMENTAL IMPACTS

Sewage treatment plants can have significant effects on the biotic status of receiving waters and can cause some water pollution, especially if the treatment process used is only basic. For example, for sewage treatment plants without nutrient removal, eutrophication of receiving water bodies can be a problem.

Water pollution (or aquatic pollution) is the contamination of water bodies, usually as a result of human activities, in such a manner that negatively affects its legitimate uses. Water pollution reduces the ability of the body of water to provide the ecosystem services that it would otherwise provide. Water bodies include for example lakes, rivers, oceans, aquifers, reservoirs and groundwater. Water pollution results when contaminants are introduced into these water bodies. Water pollution can usually be attributed to one of four sources: sewage, industry, agriculture, and urban runoff including stormwater. For example, releasing inadequately treated wastewater into natural waters can lead to degradation of these aquatic ecosystems. Water pollution can also lead to water-borne diseases for people using polluted water for drinking, bathing, washing or irrigation. Supplying clean drinking water is an important ecosystem service provided by some freshwater systems, but approximately 785 million people in the world do not have access to clean drinking water because of pollution.

7.18.1 Reuse

Fig. 11. Sludge drying beds for sewage sludge treatment at a small treatment plant at the Center for Research and Training in Sanitation, Belo Horizonte, Brazil.

7.19 IRRIGATION

Increasingly, people use treated or even untreated sewage for irrigation to produce crops. Cities provide lucrative markets for fresh produce, so are attractive to farmers. Because agriculture has to compete for increasingly scarce water resources with industry and municipal users, there is often no alternative for farmers but to use water polluted with sewage directly to water their crops. There can be significant health hazards related to using water loaded with pathogens in this way. The World Health Organization developed guidelines for safe use of wastewater in 2006. They advocate a 'multiple-barrier' approach to wastewater use, where farmers are encouraged to adopt various risk-reducing behaviors. These include ceasing irrigation a few days before harvesting to allow pathogens to die off in the sunlight, applying water carefully so it does not contaminate leaves likely to be eaten raw, cleaning vegetables with disinfectant or allowing fecal sludge used in farming to dry before being used as a human manure.

7.19.1 Reclaimed Water

Water reclamation (also called wastewater reuse, water reuse or water recycling) is the process of converting municipal wastewater (sewage) or industrial wastewater into water that can be reused for a variety of purposes. Types of reuse include: urban reuse, agricultural reuse (irrigation), environmental reuse, industrial reuse, planned potable reuse, de facto wastewater reuse (unplanned potable reuse). For example, reuse may include irrigation of gardens and agricultural fields or replenishing surface water and groundwater (i.e., groundwater recharge). Reused water may also be directed toward fulfilling certain needs in residences (e.g. toilet flushing), businesses, and industry, and could even be treated to reach drinking water standards. Treated municipal wastewater reuse for irrigation is a long-established practice, especially in arid countries. Reusing wastewater as part of sustainable water management allows water to remain as an alternative water source for human activities. This can reduce scarcity and alleviate pressures on groundwater and other natural water bodies.

7.20 GLOBAL SITUATION

Before the 20th century in Europe, sewers usually discharged into a body of water such as a river, lake, or ocean. There was no treatment, so the breakdown of the human waste was left to the ecosystem. This could lead to satisfactory results if the assimilative capacity of the ecosystem is sufficient which is nowadays not often the case due to increasing population density.

Today, the situation in urban areas of industrialized countries is usually that sewers route their contents to a sewage treatment plant rather than directly to a body of water. In many developing countries, however, the bulk of municipal and industrial wastewater is discharged to rivers and the ocean without any treatment or after preliminary treatment or primary treatment only. Doing so can lead to water pollution. Few reliable figures exist on the share of the wastewater collected in sewers that is being treated in the world. A global estimate by UNDP and UN-Habitat in 2010 was that 90% of all wastewater generated is released into the environment untreated. A more recent study in 2021 estimated that globally, about 52% of sewage is treated. However, sewage treatment rates are highly unequal for different countries around the world. For example, while high-income countries treat approximately 74% of their sewage, developing countries treat an average of just 4.2%.

The Joint Monitoring Programme (JMP) for Water Supply and Sanitation by WHO and UNICEF report in 2021 that 82% of people with sewer connections are connected to sewage treatment plants providing at least secondary treatment. However, this value varies widely between regions. For example in Europe, North America, Northern Africa and Western Asia, a total of 31 countries had universal (>99%) wastewater treatment. However, in Albania, Bermuda, North Macedonia and Serbia "less than 50% of sewered wastewater received secondary or better treatment" and in Algeria, Lebanon and Libya the value was less than 20% of sewered wastewater that wasbeing treated. The report also found that "globally, 594 million people have sewer connections that don't receive sufficient treatment. Many more are connected to wastewater treatment plants that do not provide effective treatment or comply with effluent requirements."

In October 2021, British Members of Parliament voted to continue allowing untreated sewage from combined sewer overflows to be released into waterways.

7.21 GLOBAL TARGETS

Sustainable Development Goal 6 has a Target 6.3 which is formulated as follows: "By 2030, improve water quality by reducing pollution, eliminating dumping and minimizing release of hazardous chemicals and materials, halving the proportion of untreated wastewater and substantially increasing recycling and safe reuse globally." The corresponding Indicator 6.3.1 is the "proportion of wastewater safely treated".

Data in 2020 showed that there is still too much uncollected household wastewater: Only 66% of all household wastewater flows were collected at treatment facilities in 2020 (this is determined from data from 128 countries). Based on data from 42 countries in 2015, the report stated that "32 per cent of all wastewater flows generated from point sources received at least some treatment". For sewage that has indeed been collected at centralized sewage treatment plants, about 79% went on to be safely treated in 2020.

7.22 INDUSTRIAL WASTEWATER TREATMENT PLANTS

Industrial wastewater treatment describes the processes used for treating wastewater that is produced by industries as an undesirable by-product. After treatment, the treated industrial wastewater (or effluent) may be reused or released to a sanitary sewer or to a surface water in the environment. Some industrial facilities generate wastewater that can be treated in sewage treatment plants. Most industrial processes, such as petroleum refineries, chemical and petrochemical plants have their own specialized facilities to treat their wastewaters so that the pollutant concentrations in the treated wastewater comply with the regulations regarding disposal of wastewaters into sewers or into rivers, lakes or oceans. This applies to industries that generate wastewater with high concentrations of organic matter (e.g. oil and grease), toxic pollutants (e.g. heavy metals, volatile organic compounds) or nutrients such as ammonia. Some industries install a pre-treatment system to remove some pollutants (e.g., toxic compounds), and then discharge the partially treated wastewater to the municipal sewer system.

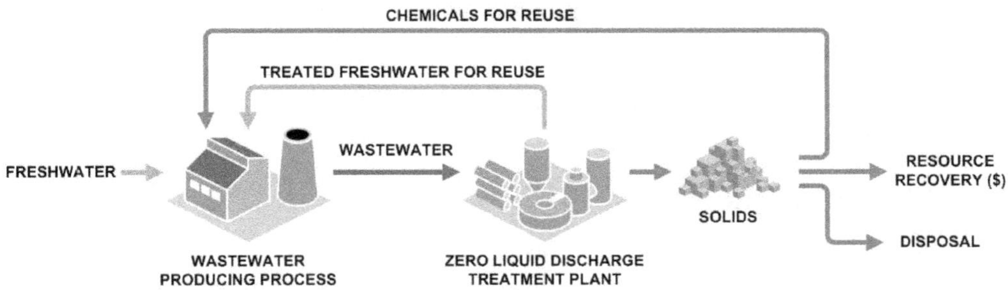

Fig. 12. Wastewater from an industrial process can be converted at a treatment plant to solids and treated water for reuse.

Most industries produce some wastewater. Recent trends have been to minimize such production or to recycle treated wastewater within the production process. Some industries have been successful at redesigning their manufacturing processes to reduce or eliminate pollutants. Sources of industrial wastewater include battery manufacturing, chemical manufacturing, electric power plants, food industry, iron and steel industry, metal working, mines and quarries, nuclear industry, oil and gas extraction, petroleum refining and petrochemicals, pharmaceutical manufacturing, pulp and paper industry, smelters, textile mills, industrial oil contamination, water treatment and wood preserving. Treatment processes include brine treatment, solids removal (e.g. chemical precipitation, filtration), oils and grease removal, removal of biodegradable organics, removal of other organics, removal of acids and alkalis, and removal of toxic materials.

7.23 TREATMENT METHODS

The various types of contamination of wastewater require a variety of strategies to remove the contamination. Most industrial processes, such as petroleum refineries, chemical and petrochemical plants have onsite facilities to treat their wastewaters so that the pollutant concentrations in the treated wastewater comply with the regulations regarding disposal of wastewaters into sewers or into rivers, lakes or oceans. Constructed wetlands are being used in an increasing number of cases as they provided high quality and productive on-site treatment. Other industrial processes that produce a lot of waste-waters such as paper and pulp production have created environmental concern, leading to development of processes to recycle water use within plants before they have to be cleaned and disposed.

An industrial wastewater treatment plant may include one or more of the following rather than the conventional treatment sequence of sewage treatment plants:

- An API oil-water separator, for removing separate phase oil from wastewater.
- A clarifier, for removing solids from wastewater.
- A roughing filter, to reduce the biochemical oxygen demand of wastewater.
- A carbon filtration plant, to remove toxic dissolved organic compounds from wastewater.
- An advanced Electrodialysis Reversal (EDR) system with ion-exchange membranes.

7.23.1 Brine Treatment

Brine treatment involves removing dissolved salt ions from the waste stream. Although similarities to seawater or brackish water desalination exist, industrial brine treatment may contain unique combinations of dissolved ions, such as hardness ions or other metals, necessitating specific processes and equipment.

Brine treatment systems are typically optimized to either reduce the volume of the final discharge for more economic disposal (as disposal costs are often based on volume) or maximize the recovery of fresh water or salts. Brine treatment systems may also be optimized to reduce electricity consumption, chemical usage, or physical footprint.

Brine treatment is commonly encountered when treating cooling tower blowdown, produced water from Steam Assisted Gravity Drainage (SAGD), produced water from natural gas extraction such as coal seam gas, frac flowback water, acid mine or acid rock drainage, reverse osmosis reject, chlor-alkali wastewater, pulp and paper mill effluent, and waste streams from food and beverage processing.

Brine treatment technologies may include: membrane filtration processes, such as reverse osmosis; ion-exchange processes such as electrodialysis or weak acid cation exchange; or evaporation processes, such as brine concentrators and crystallizers employing mechanical vapour recompression and steam. Due to the ever increasing discharge standards, there has been an emergence of the use of advance oxidation processes for the treatment of brine. Some notable examples such as Fenton's oxidation and ozonation have been employed for degradation of recalcitrant compounds in brine from industrial plants.

Reverse osmosis may not be viable for brine treatment, due to the potential for fouling caused by hardness salts or organic contaminants, or damage to the reverse osmosis membranes from hydrocarbons.

Evaporation processes are the most widespread for brine treatment as they enable the highest degree of concentration, as high as solid salt. They also produce the highest purity effluent, even distillate-quality. Evaporation processes are also more tolerant of organics, hydrocarbons, or hardness salts. However, energy consumption is high and corrosion may be an issue as the prime mover is concentrated salt water. As a result, evaporation systems typically employ titanium or duplex stainless steel materials.

7.23.2 Brine Management

Brine management examines the broader context of brine treatment and may include consideration of government policy and regulations, corporate sustainability, environmental impact, recycling, handling and transport, containment, centralized compared to on-site treatment, avoidance and reduction, technologies, and economics. Brine management shares some issues with leachate management and more general waste management. In the recent years, there has been greater prevalence in brine management due to global push for Zero liquid discharge (ZLD) /Minimal liquid discharge (MLD). In ZLD/MLD techniques, a closed water cycle is used to minimize water discharges from a system for water reuse. This concept has been gaining traction in recent years, due to increased water discharges and recent advancement in membrane technology.

7.23.3 Solids Removal

Most solids can be removed using simple sedimentation techniques with the solids recovered as slurry or sludge. Very fine solids and solids with densities close to the density of water pose special problems. In such case filtration or ultrafiltration may be required. Although flocculation may be used, using alum salts or the addition of polyelectrolytes. Wastewater from industrial food processing often requires on-site treatment before it can be discharged to prevent or reduce sewer surcharge fees. The type of industry and specific operational practices determine what types of wastewater is generated and what type of treatment is required. Reducing solids such as waste product, organic materials, and sand is often a goal of industrial wastewater treatment. Some common ways to reduce solids include primary sedimentation (clarification), Dissolved Air Flotation or (DAF), belt filtration (microscreening), and drum screening.

7.23.4 Oils and Grease Removal

The effective removal of oils and grease is dependent on the characteristics of the oil in terms of its suspension state and droplet size, which will in turn affect the choice of separator technology. Oil in industrial waste water may be free light oil, heavy oil, which tends to sink, and emulsified oil, often referred to as soluble oil. Emulsified or soluble oils will typically required "cracking" to free the oil from its emulsion. In most cases this is achieved by lowering the pH of the water matrix.

Most separator technologies will have an optimum range of oil droplet sizes that can be effectively treated.

Analyzing the oily water to determine droplet size can be performed with a video particle analyser. Each separator technology will have its own performance curve outlining optimum performance based on oil droplet size. the most common separators are gravity tanks or pits, API oil-water separators or plate packs, chemical treatment via DAFs, centrifuges, media filters and hydrocyclones.

7.24 API SEPARATORS

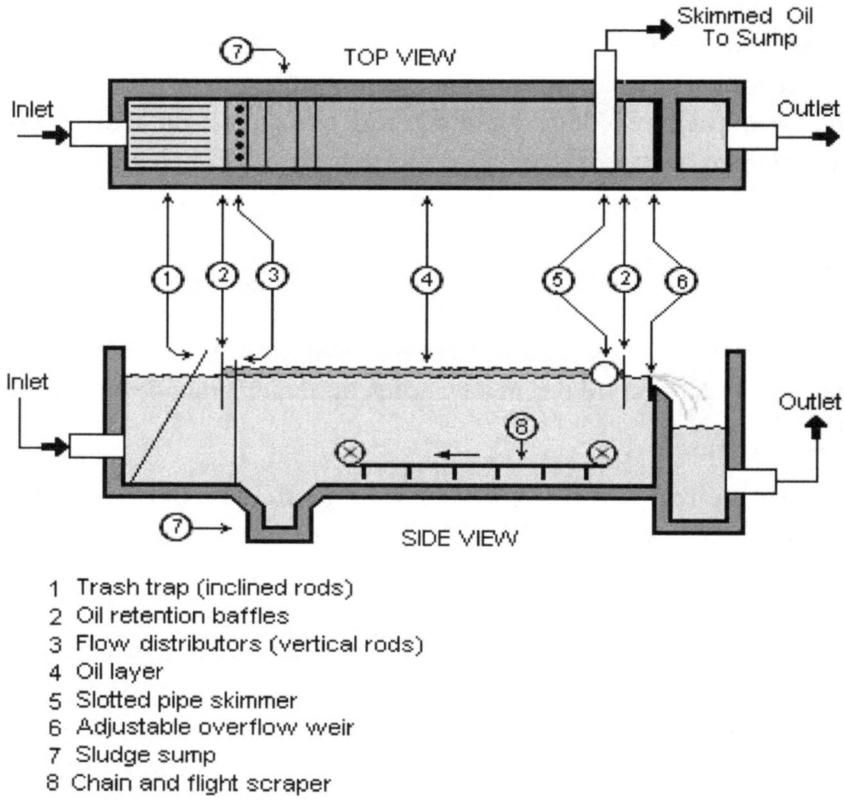

Fig. 13. A Typical API Oil-water Separator Used in Many Industries.

Many oils can be recovered from open water surfaces by skimming devices. Considered a dependable and cheap way to remove oil, grease and other hydrocarbons from water, oil skimmers can sometimes achieve the desired level of water purity. At other times, skimming is also a cost-efficient method to remove most of the oil before using membrane filters and chemical processes. Skimmers will prevent filters from blinding prematurely and keep chemical costs down because there is less oil to process.

Because grease skimming involves higher viscosity hydrocarbons, skimmers must be equipped with heaters powerful enough to keep grease fluid for discharge. If floating grease forms into solid clumps or mats, a spray bar, aerator or mechanical apparatus can be used to facilitate removal.

However, hydraulic oils and the majority of oils that have degraded to any extent will also have a soluble or emulsified component that will require further treatment to eliminate. Dissolving or emulsifying oil using surfactants or solvents usually exacerbates the problem rather than solving it, producing wastewater that is more difficult to treat.

The wastewaters from large-scale industries such as oil refineries, petrochemical plants, chemical plants, and natural gas processing plants commonly contain gross amounts of oil and suspended solids. Those industries use a device known as an API oil-water separator which is designed to separate the oil and suspended solids from their wastewater effluents. The name is derived from the fact that such separators are designed according to standards published by the American Petroleum Institute (API).

The API separator is a gravity separation device designed by using Stokes Law to define the rise velocity of oil droplets based on their density and size. The design is based on the specific gravity difference between the oil and the wastewater because that difference is much smaller than the specific gravity difference between the suspended solids and water. The suspended solids settles to the bottom of the separator as a sediment layer, the oil rises to top of the separator and the cleansed wastewater is the middle layer between the oil layer and the solids.

Typically, the oil layer is skimmed off and subsequently re-processed or disposed of, and the bottom sediment layer is removed by a chain and flight scraper (or similar device) and a sludge pump. The water layer is sent to further treatment for additional removal of any residual oil and then to some type of biological treatment unit for removal of undesirable dissolved chemical compounds.

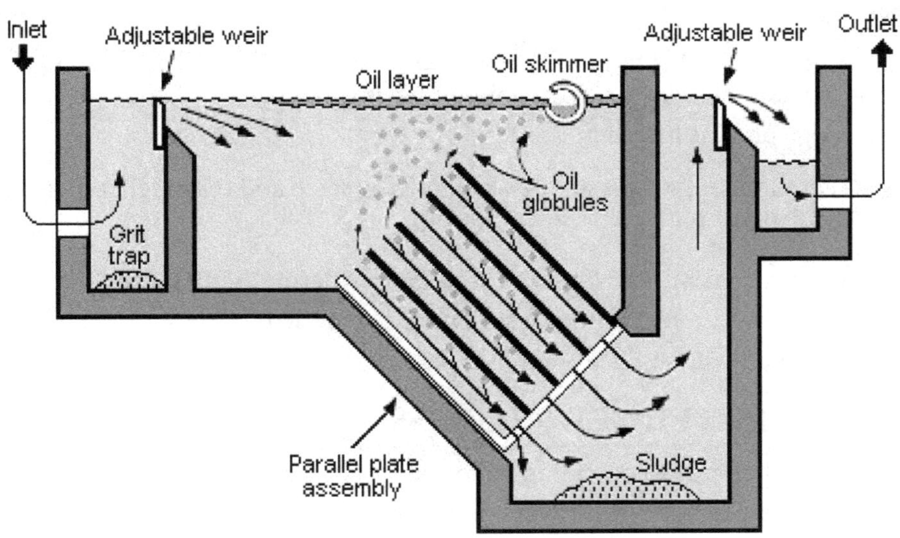

Fig. 14. A typical parallel plate separator.

Parallel plate separators are similar to API separators but they include tilted parallel plate assemblies (also known as parallel packs). The parallel plates provide more surface for suspended oil droplets to coalesce into larger globules. Such separators still depend upon the specific gravity between the suspended oil and the water. However, the parallel plates enhance the degree of oil-water separation. The result is that a parallel plate separator requires significantly less space than a conventional API separator to achieve the same degree of separation.

7.25 HYDROCYCLONE OIL SEPARATORS

Hydrocyclone oil separators operate on the process where wastewater enters the cyclone chamber and is spun under extreme centrifugal forces more than 1000 times the force of gravity. This force causes the water and oil droplets to separate. The separated oil is discharged from one end of the cyclone where treated water is discharged through the opposite end for further treatment, filtration or discharge.

7.25.1 Removal of Biodegradable Organics

Biodegradable organic material of plant or animal origin is usually possible to treat using extended conventional sewage treatment processes such as activated sludge or trickling filter. Problems can arise if the wastewater is excessively diluted with washing water or is highly concentrated such as undiluted blood or milk. The presence of cleaning agents, disinfectants, pesticides, or antibiotics can have detrimental impacts on treatment processes.

7.26 ACTIVATED SLUDGE PROCESS

Activated sludge is a biochemical process for treating sewage and industrial wastewater that uses air (or oxygen) and microorganisms to biologically oxidize organic pollutants, producing a waste sludge (or floc) containing the oxidized material. In general, an activated sludge process includes:

- An aeration tank where air (or oxygen) is injected and thoroughly mixed into the wastewater.
- A settling tank (usually referred to as a clarifier or "settler") to allow the waste sludge to settle. Part of the waste sludge is recycled to the aeration tank and the remaining waste sludge is removed for further treatment and ultimate disposal.

7.27 TRICKLING FILTER PROCESS

A trickling filter consists of a bed of rocks, gravel, slag, peat moss, or plastic media over which wastewater flows downward and contacts a layer (or film) of microbial slime covering the bed media. Aerobic conditions are maintained by forced air flowing through the bed or by natural convection of air. The process involves adsorption of organic compounds in the wastewater by the microbial slime layer, diffusion of air into the slime layer to provide the oxygen required for the biochemical oxidation of the organic compounds. The end products include carbon dioxide gas, water and other products of the oxidation. As the slime layer thickens, it becomes difficult for the air to penetrate the layer and an inner anaerobic layer is formed.

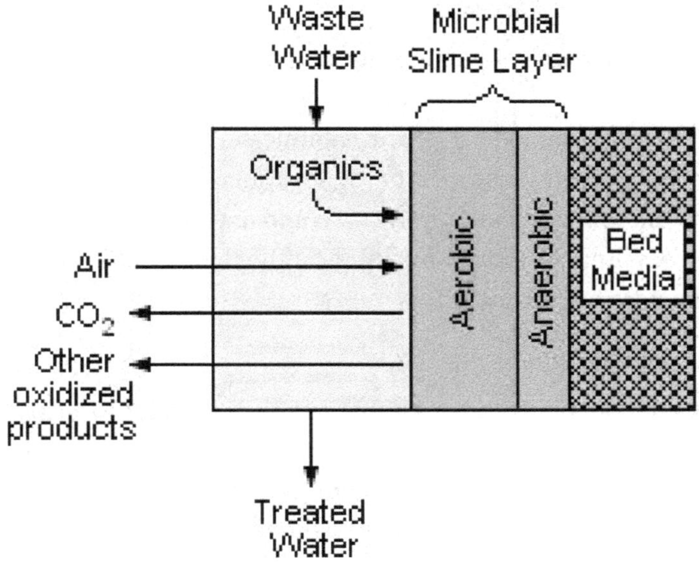

Fig. 15. A schematic cross-section of the contact face of the bed media in a trickling filter.

7.28 REMOVAL OF OTHER ORGANICS

Synthetic organic materials including solvents, paints, pharmaceuticals, pesticides, products from coke production and so forth can be very difficult to treat. Treatment methods are often specific to the material being treated. Methods include advanced oxidation processing, distillation, adsorption, ozonation, vitrification, incineration, chemical immobilisation or landfill disposal. Some materials such as some detergents may be capable of biological degradation and in such cases, a modified form of wastewater treatment can be used.

7.28.1 Removal of Acids and Alkalis

Acids and alkalis can usually be neutralised under controlled conditions. Neutralisation frequently produces a precipitate that will require treatment as a solid residue that may also be toxic. In some cases, gases may be evolved requiring treatment for the gas stream. Some other forms of treatment are usually required following neutralisation.

Waste streams rich in hardness ions as from de-ionisation processes can readily lose the hardness ions in a buildup of precipitated calcium and magnesium salts. This precipitation process can cause severe furring of pipes and can, in extreme cases, cause the blockage of disposal pipes. A 1-metre diameter industrial marine discharge pipe serving a major chemicals complex was blocked by such salts in the 1970s. Treatment is by concentration of de-ionisation waste waters and disposal to landfill or by careful pH management of the released wastewater.

7.28.2 Removal of Toxic Materials

Toxic materials including many organic materials, metals (such as zinc, silver, cadmium, thallium, etc.) acids, alkalis, non-metallic elements (such as arsenic or selenium) are generally resistant to biological processes unless very dilute. Metals can often be precipitated out by changing the pH or by treatment with other chemicals. Many, however, are resistant to treatment or mitigation and may require concentration followed by landfilling or recycling. Dissolved organics can be incinerated within the wastewater by the advanced oxidation process.

7.28.3 Smart Capsules

Molecular encapsulation is a technology that has the potential to provide a system for the recyclable removal of lead and other ions from polluted sources. Nano-, micro- and milli- capsules, with sizes in the ranges 10 nm-1μm,1μm-1mm and >1mm, respectively, are particles that have an active reagent (core) surrounded by a carrier (shell).There are three types of capsule under investigation: alginate-based capsules, carbon nanotubes, polymer swelling capsules. These capsules provide a possible means for the remediation of contaminated water.

7.28.4 Removal of Thermal Pollution

To remove heat from wastewater generated by power plants or manufacturing plants, and thus to reduce thermal pollution, the following technologies are used:

- Cooling ponds, engineered bodies of water designed for cooling by evaporation, convection, and radiation
- Cooling towers, which transfer waste heat to the atmosphere through evaporation or heat transfer
- Cogeneration, a process where waste heat is recycled for domestic or industrial heating purposes.

7.29 OTHER DISPOSAL METHODS

Some facilities such as oil and gas wells may be permitted to pump their wastewater underground through injection wells. However, wastewater injection has been linked to induced seismicity.

Agricultural wastewater treatment plants

Fig. 16 Anaerobic lagoon for treatment of dairy wastes

Agricultural wastewater treatment is a farm management agenda for controlling pollution from confined animal operations and from surface runoff that may be contaminated by chemicals in fertilizer, pesticides, animal slurry, crop residues or irrigation water. Agricultural wastewater treatment is required for continuous confined animal operations like milk and egg production. It may be performed in plants using mechanized treatment units similar to those used for industrial wastewater. Where land is available for ponds, settling basins and facultative lagoons may have lower operational costs for seasonal use conditions from breeding or harvest cycles. Animal slurries are usually treated by containment in anaerobic lagoons before disposal by spray or trickle application to grassland. Constructed wetlands are sometimes used to facilitate treatment of animal wastes.

Nonpoint source pollution includes sediment runoff, nutrient runoff and pesticides. Point source pollution includes animal wastes, silage liquor, milking parlour (dairy farming) wastes, slaughtering waste, vegetable washing water and firewater. Many farms generate nonpoint source pollution from surface runoff which is not controlled through a treatment plant.

7.30 LEACHATE TREATMENT PLANTS

Leachate treatment plants are used to treat leachate from landfills. Treatment options include: biological treatment, mechanical treatment by ultrafiltration, treatment with active carbon filters, electrochemical treatment including electrocoagulation by various proprietary technologies and reverse osmosis membrane filtration using disc tube module technology.

7.30.1 Unit Processes

The unit processes involved in wastewater treatment include physical processes such as settlement or flotation and biological processes such oxidation or anaerobic treatment. Some wastewaters require specialized treatment methods. At the simplest level, treatment of most wastewaters is carried out through separation of solids from liquids, usually by sedimentation. By progressively converting dissolved material into solids, usually a biological floc or biofilm, which is then settled out or separated, an effluent stream of increasing purity is produced.

7.30.2 Phase Separation

Phase separation transfers impurities into a non-aqueous phase. Phase separation may occur at intermediate points in a treatment sequence to remove solids generated during oxidation or polishing. Grease and oil may be recovered for fuel or saponification. Solids often require dewatering of sludge in a wastewater treatment plant. Disposal options for dried solids vary with the type and concentration of impurities removed from water.

7.30.3 Sedimentation

Solids such as stones, grit, and sand may be removed from wastewater by gravity when density differences are sufficient to overcome dispersion by turbulence. This is typically achieved using a grit channel designed to produce an optimum flow rate that allows grit to settle and other less-dense solids to be carried forward to the next treatment stage. Gravity separation of solids is the primary treatment of sewage, where the unit process is called "primary settling tanks" or "primary sedimentation tanks." It is also widely used for the treatment of other types of wastewater. Solids that are denser than water will accumulate at the bottom of quiescent settling basins. More complex clarifiers also have skimmers to simultaneously remove floating grease such as soap scum and solids such as feathers, wood chips, or condoms. Containers like the API oil-water separator are specifically designed to separate non-polar liquids.

7.31 BIOLOGICAL AND CHEMICAL PROCESSES

7.31.1 Oxidation

Oxidation reduces the biochemical oxygen demand of wastewater, and may reduce the toxicity of some impurities. Secondary treatment converts organic compounds into carbon dioxide, water, and biosolids through oxidation and reduction reactions. Chemical oxidation is widely used for disinfection.

7.31.2 Biochemical Oxidation (Secondary Treatment)

Secondary treatment is the removal of biodegradable organic matter (in solution or suspension) from sewage or similar kinds of wastewater. The aim is to achieve a certain degree of effluent quality in a sewage treatment plant suitable for the intended disposal or reuse option. A "primary treatment" step often precedes secondary treatment, whereby physical phase separation is used to remove settleable solids. During secondary treatment, biological processes are used to remove dissolved and suspended organic matter measured as biochemical oxygen demand (BOD). These processes are performed by microorganisms in a managed aerobic or anaerobic process depending on the treatment technology. Bacteria and protozoa consume biodegradable soluble organic contaminants (e.g. sugars, fats, and organic short-chain carbon molecules from human waste, food waste, soaps and detergent) while reproducing to form cells of biological solids. Secondary treatment is widely used in sewage treatment and is also applicable to many agricultural and industrial wastewaters.

7.31.3 Chemical Oxidation

Advanced oxidation processes are used to remove some persistent organic pollutants and concentrations remaining after biochemical oxidation. Disinfection by chemical oxidation kills bacteria and microbial pathogens by adding hydroxyl radicals such as ozone, chlorine or hypochlorite to wastewater. These hydroxyl radical then break down complex compounds in the organic pollutants into simple compounds such as water, carbon dioxide, and salts.

8

RAIN WATER HARVESTING

8.1 OBJECTIVES

After studying this chapter you will be able to understand:

- Rain water harvesting
- Comparison with peak oil.
- Chlorine sterilisation
- Water conservation.
- Clemmate change.
- Moisture conservation.

8.2 INTRODUCTION

The term rainwater harvesting is being frequently used these days, however, the concept of water harvesting is not new for India. Water harvesting techniques had been evolved and developed centuries ago. Ground water resource gets naturally recharged through percolation. But due to indiscriminate development and rapid urbainzation, exposed surface for soil has been reduced drastically with resultant reduction in percolation of rainwater, thereby depleting ground water resource.

Rainwater harvesting is the process of augmenting the natural filtration of rainwater in to the underground formation by some artificial methods. "Conscious collection and storage of rainwater to cater to demands of water, for drinking, domestic purpose & irrigation is termed as Rainwater Harvesting. "

This is perhaps one of the most frequently asked question, as to why one should harvest rainwater. There are many reasons but following are some of the important ones.

- To arrest ground water decline and augment ground water table

- To beneficiate water quality in aquifers
- To conserve surface water runoff during monsoon
- To reduce soil erosion
- To inculcate a culture of water conservation

How to harvest rainwater: Broadly there are two ways of harvesting rainwater:

(i) Surface runoff harvesting

(ii) Roof top rainwater harvesting

Surface runoff harvesting: In urban area rainwater flows away as surface runoff. This runoff could be caught and used for recharging aquifers by adopting appropriate methods.

Roof top rainwater harvesting: It is a system of catching rainwater where it falls. In rooftop harvesting, the roof becomes the catchments, and the rainwater is collected from the roof of the house/building. It can either be stored in a tank or diverted to artificial recharge system. This method is less expensive and very effective and if implemented properly helps in augmenting the ground water level of the area.

The first stage in this process, before treatment can begin, is the harvesting and collection of rainwater. There are many ways to collect rainwater, from dedicated dams to roof runoff gutter systems connected to water tanks. Preliminary stages of water treatment can be introduced at this point, with filters in place to prevent leaves and debris from being collected along with the water.

Two primary sanitisation options are available for rainwater treatment:

- **Chlorine Treatment:** This is an excellent solution when large amounts of water will be stored for later use and in general industrial applications.
- **Ultraviolet Light Treatment:** This treatment is commonly used when smaller amounts of stored water are required for a shorter time period and is a great alternative for industrial applications sensitive to corrosion from chlorine exposure.

Businesses located in regions that experience sporadic to heavy rainfall can significantly reduce their water expenses by implementing a rainwater harvesting system. Rainwater treatment and sterilisation are both necessary for making rainwater compliant with regulations and safe to use for your business.

Although rainwater appears to be clean to the naked eye, it is not safe to use until treated as it poses a risk to staff exposed to it. This pathogen-rich rainwater can cause sickness or infections if it comes into contact with open wounds or ingested.

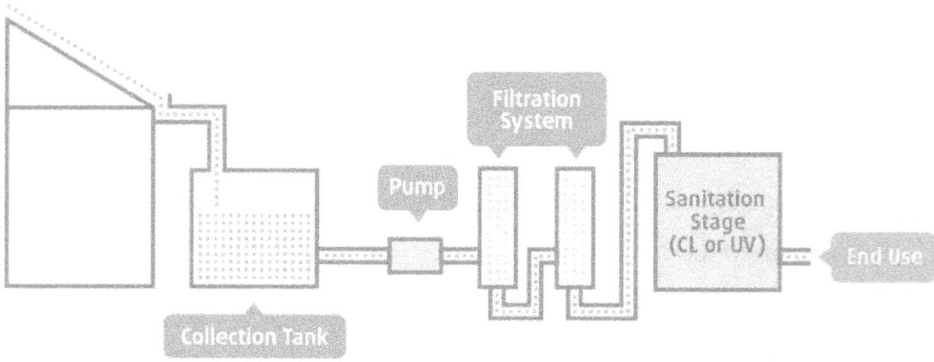

Fig. 1. Cleanawater supplies packaged, low maintenance rainwater harvesting solutions to suit all inground and above ground rainwater harvesting applications.

8.3 SYSTEM SETUP

Rainwater harvesting systems can range in complexity, from systems that can be installed with minimal skills, to automated systems that require advanced setup and installation. The basic Rainwater harvesting system is more of a plumbing job than a technical job as all the outlets from the building terrace are connected through a pipe to an underground tank that stores water.

Systems are ideally sized to meet the water demand throughout the dry season since it must be big enough to support daily water consumption. Specifically, the rainfall capturing area such as a building roof must be large enough to maintain adequate flow. The water storage tank size should be large enough to contain the captured water.

For low-tech systems, there are many low-tech methods used to capture rainwater: rooftop systems, surface water capture, and pumping the rainwater that has already soaked into the ground or captured in reservoirs and storing it into tanks (cisterns).

Before a rainwater harvesting system is built, it's helpful to use digital tools. For instance, if you want to detect if a region has a high rainwater harvesting potential, rainwater harvesting GIS maps can be made using an online interactive tool. Or if you need to estimate how much water is needed to fulfill a community's water needs, the Rain is Gain tool helps with this. Tools like these can save time and money before a commitment to build a system is undertaken, in addition to making the project sustainable and last a long time.

8.3.1 Life Cycle Assessment: Design for Environment

Contemporary system designs require an analysis of not only the economic and technical performance of a system, but also the environmental performance. Life Cycle Assessment is a methodology used to evaluate the environmental impacts of a precut or systems, from cradle-to-grave of its' lifetime. Devkota et al., developed such a methodology for rainwater harvesting, and found that the building design (e.g., dimensions) and function (e.g., educational, residential, etc.) play critical roles in the environmental performance of the system. The Economic and Environmental Analysis of Sanitations Technologies, EEAST model evaluates the greenhouse gas emissions and cost of such systems over the lifetime of a variety of building types.

To address the functional parameters of rainwater harvesting systems, a new metric was developed - the demand to supply ratio (D/S) - identifying the ideal building design (supply) and function (demand) in regard to the environmental performance of rainwater harvesting for toilet flushing. With the idea that supply of rainwater not only saves the potable water, but also saves the stormwater entering the combined sewer network (thereby requiring treatment), the savings in environmental emissions were higher if the buildings are connected to a combined sewer network compared to separate one.

8.4 RAIN WATER HARVESTING BY FRESHWATER FLOODED FORESTS

Rain water harvesting is possible by growing fresh water flooded forests without losing the income from the used /submerged land. The main purpose of the rain water harvesting is to utilize the locally available rain water to meet water requirements throughout the year without the need of huge capital expenditure. This would facilitate availability of uncontaminated water for domestic, industrial and irrigation needs.

8.4.1 New Approaches

Instead of using the roof for catchment, the RainSaucer, which looks like an upside down umbrella, collects rain straight from the sky. This decreases the potential for contamination and makes potable water for developing countries a potential application. Other applications of this free standing rainwater collection approach are sustainable gardening and small plot farming.

A Dutch invention called the GroasisWaterboxx is also useful for growing trees with harvested and stored dew and rainwater.

Traditionally, storm water management using detention basins served a single purpose. However, Optimized Real-Time Control (OptiRTC) lets this infrastructure double as a source of rainwater harvesting without compromising the existing detention capacity. This has been used in the EPA headquarters to evacuate stored water prior to storm events, thus reducing wet weather flow while ensuring water availability for later reuse. This has the benefit of increasing water quality released and decreasing the volume of water released during combined sewer overflow events.

Generally, check dams are constructed across the streams to enhance the percolation of surface water in to the sub soil strata. The water percolation in the water impounded area of the check dams, can be enhanced artificially many folds by loosening the sub soil strata / overburden by using ANFO explosives as used in open cast mining. Thus local aquifers can be recharged quickly by using the available surface water fully for using in the dry season.

8.4.2 History of Rain Water Harvesting

Around the third century BC, the farming communities in Balochistan (now located in Pakistan, Afghanistan and Iran), and Kutch, India, used rainwater harvesting for agriculture and many uses also. In ancient Tamil Nadu, rainwater harvesting was done by Chola kings. Rainwater from the Brihadeeswarar temple (located in Balaganpathy Nagar, Thanjavur, India) was collected in Shivaganga tank. During the later Chola period, the Vîrânam tank was built (1011 to 1037 CE) in Cuddalore district of Tamil Nadu to store water for drinking and irrigation purposes. Vîrânam is a 16 km (9.9 mi) long tank with a storage capacity of 1,465,000,000 cubic feet (41,500,000 m).

Rainwater harvesting was done in the Indian states of Madhya Pradesh, Maharashtra, and Chhattisgarh in the olden days. Ratanpur, in the state of Chhattisgarh, had around 150 ponds. Most of the tanks or ponds were utilized in agriculture works.

8.4.3 Present Day

- Currently in China and Brazil rooftop rainwater harvesting is being practiced for providing drinking water, domestic water, water for livestock, water for small irrigation and a way to replenish groundwater levels. Gansu province in China and semi-arid north east Brazil have the largest rooftop rainwater harvesting projects ongoing.
- In Bermuda, the law requires all new construction to include rainwater harvesting adequate for the residents.
- The U.S. Virgin Islands have a similar law.
- In Senegal and Guinea-Bissau, the houses of the Diola-people are frequently equipped with homebrew rainwater harvesters made from local, organic materials.
- In the Irrawaddy Delta of Myanmar, the groundwater is saline and communities rely on mud-lined rainwater ponds to meet their drinking water needs throughout the dry season. Some of these ponds are centuries old and are treated with great reverence and respect.
- In the United States: until 2009 in Colorado, water rights laws almost completely restricted rainwater harvesting; a property owner who captured rainwater was deemed to be stealing it from those who have rights to take water from the watershed. Now, residential well owners that meet certain criteria may obtain a permit to install a rooftop precipitation collection system (SB 09-080). Up to 10 large scale pilot studies may also be permitted (HB 09-1129). The main factor in persuading the Colorado Legislature to change the law was a 2007 study that

found that in an average year, 97% of the precipitation that fell in Douglas County, in the southern suburbs of Denver, never reached a stream—it was used by plants or evaporated on the ground. In Colorado you cannot even drill a water well unless you have at least 35 acres (14 ha). In New Mexico, rainwater catchment is mandatory for new dwellings in Santa Fe. Texas offers a sales tax exemption on the purchase of rainwater harvesting equipment. Both Texas and Ohio allow the practice even for potable purposes. Oklahoma passed the Water for 2060 Act in 2012, to promote pilot projects for rainwater and graywater use among other water saving techniques.

- In Beijing, some housing societies are now adding rain water in their main water sources after proper treatment.
- In Ireland, Professor MichealMcginley established a project to design a rain water harvesting prototype in the Biosystems design Challenge Module at University College Dublin

8.4.4 Canada

A number of Canadians have started implementing rainwater harvesting systems for use in stormwater reduction, irrigation, laundry, and lavatory plumbing. Substantial reform to Canadian law since the mid 2000s has increased the use of this technology in agricultural, industrial, and residential use; but ambiguity remains amongst legislation in many provinces. Bylaws and local municipal codes often regulate rainwater harvesting.

8.4.5 India

- **Karnataka:** In Bangalore it is mandatory for adoption of rain water harvesting for every owner or the occupier of a building having the sital area measuring 60 ft (18.3 m) X 40 ft (12.2 m) and above and for newly constructed building measuring 30 ft (9.1 m) X 40 ft (12.2 m) and above dimension. In this regard BWSSB has initiated and constructed "Rain Water Harvesting Theme Park" in the name of Sir. M. Visvesvaraya in 1.2 acres (4,900 m) land situated at Jayanagar, Bangalore. In this park 26 different type of Rain Water Harvesting models are demonstrated along with the water conservation tips. The Auditorium on the first floor is set up with Green Air conditioning system and the same will be utilized to arrange the meeting and showing the video clip about the rain water harvesting to students as well as general public.
- **Tamil Nadu:** In the state of Tamil Nadu, rainwater harvesting was made compulsory for every building to avoid groundwater depletion. It gave excellent results within five years, and every state took it as role model. Since its implementation, Chennai saw a 50 percent rise in water level in five years and the water quality significantly improved.

- **Rajasthan:** In Rajasthan, rainwater harvesting has traditionally been practiced by the people of the Thar Desert. There are many ancient water harvesting systems in Rajasthan, which have now been revived. Water harvesting systems are widely used in other areas of Rajasthan as well, for example the chauka system from the Jaipur district.
- **Maharashtra:** At present, in Pune, rainwater harvesting is compulsory for any new housing society to be registered.
- In Mumbai city in Maharashtra rain water harvesting is being considered as a good solution to solve water crisis.
- The Mumbai city council is planning to make rainwater harvesting mandatory for large societies.
- An attempt has been made at the Department of Chemical Engineering, IISc, Bangalore to harvest rainwater using upper surface of a solar still, which was used for water distillation

8.4.6 Israel

- The Southwest Center for the Study of Hospital and Healthcare Systems in cooperation with Rotary International is sponsoring a rainwater harvesting model program across the country. The first rainwater catchment system was installed at an elementary school in Lod, Israel. The project is looking to expand to Haifa in its third phase. The Southwest Center has also partnered with the Water Resources Action Project (WRAP) of Washington D.C. WRAP currently has rainwater harvesting projects in the West Bank.
- Rainwater harvesting systems are being installed in local schools for the purpose of educating schoolchildren about water conservation principles and bridging divides between people of different religious and ethnic backgrounds all while addressing the water scarcity issue that the Middle East faces.

8.4.7 Sri Lanka

- Rainwater harvesting has been a popular method of obtaining water for agriculture and for drinking purposes in rural homes.
- The legislation to promote rainwater harvesting was enacted through the Urban Development Authority (Amendment) Act, No. 36 of 2007.
- Lanka rainwater harvesting forum is leading the Sri Lanka's initiative.

8.4.8 South Africa

- The South African Water Research Commission has supported research into rainwater harvesting. Reports on this research are available on their 'Knowledge Hub'.

- Studies in arid, semi-arid and humid regions have confirmed that techniques such as mulching, pitting, ridging and modified run-on plots are effective for small-scale crop production.

8.4.9 United Kingdom

In the United Kingdom, water butts are often found in domestic gardens to collect rainwater, which is then used to water the garden. However, the British government's Code For Sustainable Homes encourages fitting large underground tanks to new-build homes to collect rainwater for flushing toilets, washing clothes, watering the garden, and washing cars. This reduces by 50 percent the amount of mains water used by the home.

8.4.10 United States

- **Colorado:** Until 2009, water rights laws almost completely restricted rainwater harvesting; a property owner who captured rainwater was deemed to be stealing it from those who have rights to take water from the watershed. Now, residential well owners that meet certain criteria may obtain a permit to install a rooftop precipitation collection system (SB 09-080). Up to 10 large scale pilot studies may also be permitted (HB 09-1129). The main factor in persuading the Colorado Legislature to change the law was a 2007 study that found that in an average year, 97% of the precipitation that fell in Douglas County, in the southern suburbs of Denver, never reached a stream—it was used by plants or evaporated on the ground. In Colorado you cannot even drill a water well unless you have at least 35 acres (14 ha)
- **New Mexico:** In New Mexico, rainwater catchment is mandatory for new dwellings in Santa Fe.
- Texas offers a sales tax exemption on the purchase of rainwater harvesting equipment. Both Texas and Ohio allow the practice even for potable purposes.
- Oklahoma passed the Water for 2060 Act in 2012, to promote pilot projects for rainwater and graywater use among other water saving techniques.

8.5 PEAK WATER

Peak water is a concept that underlines the growing constraints on the availability, quality, and use of freshwater resources.

Peak water is defined in a 2010 peer-reviewed article in the Proceedings of the National Academy of Sciences by Peter Gleick and MeenaPalaniappan. They distinguish between peak renewable, peak non-renewable, and peak ecological water in order to demonstrate the fact that although there is a vast amount of water on the planet, sustainably managed water is becoming scarce.

Lester R. Brown, president of the Earth Policy Institute, wrote in 2013 that although there was extensive literature on peak oil, it was peak water that is "the real threat to our future". An assessment was published in August 2011 in the Stockholm International Water Institute's journal. Much of the world's water in underground aquifers and in lakes can be depleted and thus resembles a finite resource. The phrase peak water sparks debates similar to those about peak oil. In 2010, New York Times chose "peak water" as one of its 33 "Words of the Year".

There are concerns about impending peak water in several areas around the world:

- Peak renewable water, where entire renewable flows are being consumed for human use
- Peak non-renewable water, where groundwater aquifers are being overpumped (or contaminated) faster than nature recharges them (this example is most like the peak oil debate)
- Peak ecological water, where ecological and environmental constraints are overwhelming the economic benefits provided by water use

If present trends continue, 1.8 billion people will be living with absolute water scarcity by 2025, and two-thirds of the world could be subject to water stress. Ultimately, peak water is not about running out of freshwater, but about reaching physical, economic, and environmental limits on meeting human demands for water and the subsequent decline of water availability and use.

8.6 COMPARISON WITH PEAK OIL

The Hubbert curve has become popular in the scientific community for predicting the depletion of various natural resources. M. King Hubbert created this measurement device in 1956 for a variety of finite resources such as coal, oil, natural gas and uranium.Hubbert's curve was not applied to resources such as water originally, since water is a renewable resource. Some forms of water, however, such as fossil water, exhibit similar characteristics to oil, and overpumping (faster than the rate of natural recharge of groundwater) often results in a Hubbert-type peak. A modified Hubbert curve applies to any resource that can be harvested faster than it can be replaced. Like peak oil, peak water is inevitable given the rate of extraction of certain water systems. A current argument is that growing populations and demands for water will inevitably lead to non-renewable use of water resources.

8.6.1 Water Supply

Fresh water is a renewable resource, yet the world's supply of clean, fresh water is under increasing demand for human activities. The world has an estimated 1.34 billion cubic kilometers of water, but 96.5% of it is salty. Almost 70% of fresh water can be found in the ice caps of Antarctica and Greenland. Less than 1% of this water on Earth is accessible to humans, the rest is contained in soil moisture or deep underground. Accessible freshwater is located in lakes, rivers, reservoirs and shallow underground sources. Rainwater and snowfall do very little to replenish these stocks of freshwater.The amount of available freshwater supply in some regions is decreasing because of (i) climate change, which has caused receding glaciers, reduced stream and river flow, and shrinking lakes; (ii) contamination of water by human and industrial wastes; and (iii) overuse of non-renewable groundwater aquifers. Many aquifers have been over-pumped and are not recharging quickly. Although the total freshwater supply is not used up, much has become polluted, salted, unsuitable or otherwise unavailable for drinking, industry and agriculture.

8.6.2 Water Demand

Water demand already exceeds supply in many parts of the world, and as the world population continues to rise, many more areas are expected to experience this imbalance in the near future. Agriculture represents 70% of freshwater use worldwide.

Agriculture, industrialization and urbanization all serve to increase water consumption.

8.7 FRESHWATER WITHDRAWAL BY COUNTRY

The highest total annual consumption of water comes from India, China and the United States, countries with large populations, extensive agricultural irrigation, and demand for food.

8.7.1 India

India has 20 percent of the Earth's population, but only four percent of its water. Water tables are dropping rapidly in some of India's main agricultural areas. The Indus and Ganges rivers are tapped so heavily that, except in rare wet years, they no longer reach the sea.

India has the largest water withdrawal out of all the countries in the world. Eighty-six percent of that water supports agriculture. That heavy use is dictated in large part by what people eat. People in India consume a lot of rice. Rice farmers in India typically get less than half the yield per unit area while using ten times more water than their Chinese counterparts. Economic development can make things worse because as people's living standards rise, they tend to eat more meat, which requires lots of water to produce. Growing a tonne of grain requires 1,000 tonnes of water; producing a tonne of beef requires 15,000 tonnes. To make a single hamburger requires around 4,940 liters (1,300 gallons) of water A glass of orange juice needs 850 liters (225 gallons) of freshwater to produce.

8.7.2 China

China, the world's most populous country, has the second largest water withdrawal; 68% supports agriculture while its growing industrial base consumes 26%. China is facing a water crisis where water resources are overallocated, used inefficiently, and severely polluted by human and industrial wastes. One-third of China's population lacks access to safe drinking water. Rivers and lakes are dead and dying, groundwater aquifers are over-pumped, uncounted species of aquatic life have been driven to extinction, and direct adverse impacts on both human and ecosystem health are widespread and growing.

In western China's Qinghai province, through which the Yellow River's main stream flows, more than 2,000 lakes have disappeared over the last 20 years. There were once 4,077 lakes. Global climate change is responsible for the reduction in flow of the (Huang He) Yellow River over the past several decades. The source of the Yellow River is the Qinghai-Xizang Tibetan Plateau where the glaciers are receding sharply.

In Hebei Province, which surrounds Beijing, the situation is much worse. Hebei is one of China's major wheat and corn growing provinces. The water tables have been falling fast throughout Hebei. The region has lost 969 of its 1,052 lakes. About 500,000 people are affected by a shortage of drinking water due to continuing droughts. Hydro-power generation is also impacted. Beijing and Tianjin depend on Hebei Province to supply their water from the Yangtze River. Beijing gets its water via the newly constructed South-North Water Transfer Project. The river originates in a glacier on the eastern part of the Tibetan Plateau.

8.7.3 United States

The United States has about 5% of the world's population, yet it uses almost as much water as India (~1/5 of world) or China (1/5 of world) because substantial amounts of water are used to grow food exported to the rest of the world. The U.S. agricultural sector consumes more water than the industrial sector, though substantial quantities of water are withdrawn (but not consumed) for power plant cooling systems. 40 out of 50 state water managers expect some degree of water stress in their state in the next 10 years.

The Ogallala Aquifer in the southern high plains (Texas and New Mexico) is being mined at a rate that far exceeds replenishment—a classic example of peak non-renewable water. Portions of the aquifer will not naturally recharge due to layers of clay between the surface and the water-bearing formation, and because rainfall rates simply do not match rates of extraction for irrigation. The term fossil water is sometimes used to describe water in aquifers that was stored over centuries to millennia. Use of this water is not sustainable when the recharge rate is slower than the rate of groundwater extraction.

In California, massive amounts of groundwater are also being withdrawn from Central Valley groundwater aquifers—unreported, unmonitored, and unregulated. California's Central Valley is home to one-sixth of all irrigated land in the United States, and the state leads the nation in agricultural production and exports. The inability to sustain groundwater withdrawals over time may lead to adverse impacts on the region's agricultural productivity.

The Central Arizona Project (CAP) is a 336-mile (541 km) long canal that diverts 489 billion US gallons (1.85×10 m) a year from the Colorado River to irrigate more than

300,000 acres (1,200 km) of farmland. The CAP project also provides drinking water for Phoenix and Tucson. It has been estimated that Lake Mead, which dams the Colorado, has a 50-50 chance of running dry by 2021.

The Ipswich River near Boston now runs dry in some years due to heavy pumping of groundwater for irrigation. Maryland, Virginia and the District of Columbia have been fighting over the Potomac River. In drought years like 1999 or 2003, and on hot summer days the region consumes up to 85 percent of the river's flow.

8.8 PER CAPITA WITHDRAWAL OF WATER

8.8.1 Turkmenistan

Turkmenistan gets most of its water from the Amu Darya River. The Qaraqum Canal is a canal system that takes water from the Amu Darya River and distributes the water out over the desert for irrigation of its orchard crops and cotton. Turkmenistan uses the most water per capita in the world because only 55% of the water delivered to the fields actually reaches the crops.

8.8.2 Kazakhstan and Uzbekistan

The two rivers feeding the Aral Sea were dammed up and the water was diverted to irrigate the desert so that cotton could be produced. As a result, the Aral Sea's water has become much saltier and the sea's water level has decreased by over 60%. Drinking water is now contaminated with pesticides and other agricultural chemicals and contains bacteria and viruses. The climate has become more extreme in the area surrounding it.

8.8.3 Saudi Arabia

According to Walid A. Abderrahman (2001), "Water Demand Management in Saudi Arabia", Saudi Arabia reached peak water in the early 1990s, at more than 30 billion cubic meters per year, and declined afterward. The peak had arrived at about midpoint, as expected for a Hubbert curve. Today, the water production is about half the peak rate. Saudi Arabian food production has been based on "fossil water"—water from ancient aquifers that is being recharged very slowly, if at all. Like oil, fossil water is non-renewable, and it is bound to run out someday. Saudi Arabia has abandoned its self-sufficient food production and is now importing virtually all of its food. Saudi Arabia has built desalination plants to provide about half the country's freshwater. The remainder comes from groundwater (40%), surface water (9%) and reclaimed wastewater (1%).

8.8.4 Libya

Libya is working on a network of water pipelines to import water, called the Great Manmade River. It carries water from wells tapping fossil water in the Sahara desert to the cities of Tripoli, Benghazi, Sirte and others. Their water also comes from desalination plants.

8.8.5 Yemen

Peak water has occurred in Yemen. Sustainability is no longer attainable in Yemen, according to the government's five-year water plan for 2005–2009. The aquifer that supplies Sana'a, the capital of Yemen, will be depleted by 2009. In its search for water in the basin, the Yemeni government has drilled test wells that are 2 kilometers (1.2 mi) deep, depths normally associated with the oil industry, but it has failed to find water. Yemen must soon choose between relocating the city and building a pipeline to coastal desalination plants. The pipeline option is complicated by Sana'a's altitude of 2,250 m (7,380 ft).

As of 2010, the threat of running out of water was considered greater than that of Al-Qaeda or instability. There was speculation that Yemenis would have to abandon mountain cities, including Sana'a, and move to the coast. The cultivation of khat and poor water regulation by the government were partly blamed.

8.8.6 United Arab Emirates

United Arab Emirates has a rapidly growing economy and very little water to support it. UAE requires more water than is naturally available. They have reached peak water. To solve this, UAE has a desalination plant near Ruwais and ships its water via pipeline to Abu Dhabi.

8.9 CONSEQUENCES

8.9.1 Famine

Water shortage may cause famine in Pakistan. Pakistan has approximately 35 million acres (140,000 km) of arable land irrigated by canals and tube wells, mostly using water from the Indus River. Dams were constructed at Chashma, Mangla, and Tarbela to feed the irrigation system. Since the completion of the Tarbela Dam in 1976 no new capacity has been added despite astronomical growth in population. The gross capacity of the three dams has decreased because of sedimentation, a continual process. Per-capita surface-water availability for irrigation was 5,260 cubic meters per year in 1951. This has been reduced to a mere 1,100 cubic meters per year in 2006.

8.9.2 Health Problems

The quality of drinking water is vital for human health. Peak water constraints result in people not having access to safe water for basic personal hygiene. "Infectious waterborne diseases such as diarrhea, typhoid, and cholera are responsible for 80 percent of illnesses and deaths in the developing world, many of them children. One child dies every eight seconds from a waterborne disease; 15 million children a year."

Vital aquifers everywhere are becoming contaminated with toxins. Once an aquifer is contaminated, it is not likely that it can ever recover. Contaminants are more likely to cause chronic health effects. Water can be contaminated from pathogens such as bacteria, viruses, and parasites. Also, toxic organic chemicals can be a source of water contamination. Inorganic contaminants include toxic metals like arsenic, barium, chromium, lead, mercury, and silver. Nitrates are another source of inorganic contamination. Finally, leaching radioactive elements into the water supply can contaminate it.

8.9.3 Human Conflicts Over Water

Some conflicts of the future may be fought over the availability, quality, and control of water. Water has also been used as a tool in conflicts or as a target during conflicts that start for other reasons. Water shortages may well result in water conflicts over this precious resource.

In West Africa and other places like Nepal, Bangladesh, India (such as the Ganges Delta), and Peru, major changes in the rivers generate a significant risk of violent conflict in coming years. Water management and control could play a part in future resource wars over scarce resources.

8.10 SOLUTIONS

Freshwater usage has great potential for better conservation and management as it is used inefficiently nearly everywhere, but until actual scarcity hits, people tend to take access to freshwater for granted.

8.10.1 Treatment Process

Rainwater may contain pollution, animal excrement and other particles which are harmful to humans, plants and animals. Therefore collected rainwater needs to be treated before it is safe to use for business.

There are two primary steps to rainwater treatment: filtration and sterilisation using chlorine or UV light.

8.10.2 Filtration

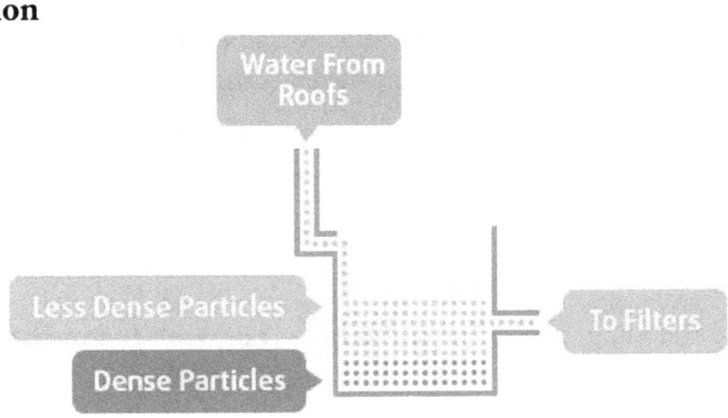

Fig. 2. The first step of rainwater treatment involves removing sediment and small particles.

Rainwater harvested from roofs travels down a pipe into a storage tank or pit. Dense sediment will settle at the bottom of the tank. A float switch triggers the operation of the system to commence pumping or a pressure drop inline indicates the need to treat and transfer water instantaneously.

Filtration consists of removing 1-5 micron solids via a cartridge, screen or sand filter prior to sanitisation. Filters also have the option to be self cleaning.

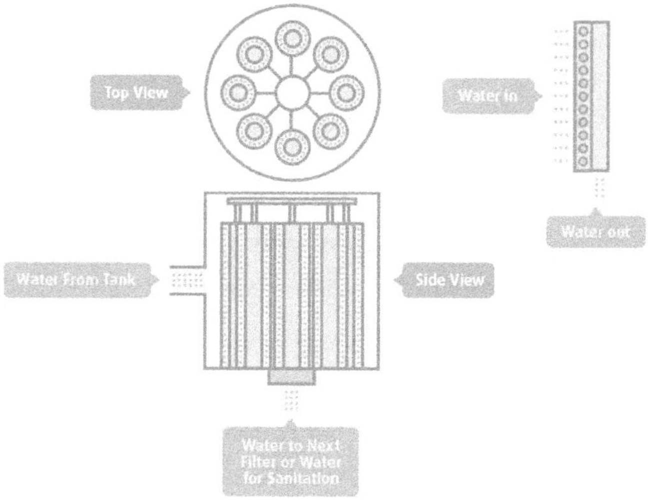

Fig. 3. Filtration.

8.10.3 Chlorine Sterilisation

Adding chlorine to water is a simple and effective way to sterilise filtered rainwater.

The Victorian Government Department of Human Services recommends the following chlorine dosage and concentration:

- The initial dosage to be a minimum of 5 milligrams of chlorine per litre of rainwater to be treated.
- A concentration of 0.5 milligrams of chlorine for every 1 litre of rainwater must be maintained, 30 minutes after the initial dose.

Therefore for a 50L tank of rainwater, a total of 250mg of chlorine is needed for sterilisation. Then 30 minutes after the initial dose, the concentration must be maintained at 25mg of chlorine per litre.

Advantages of Chlorine Sterilisation

- Chlorine treatment is an ideal solution for rainwater that needs to be stored for future use.
- Rainwater is compliant with the treatment regulations at all times, as a sampling pump will automatically regulate the appropriate chlorine concentration.

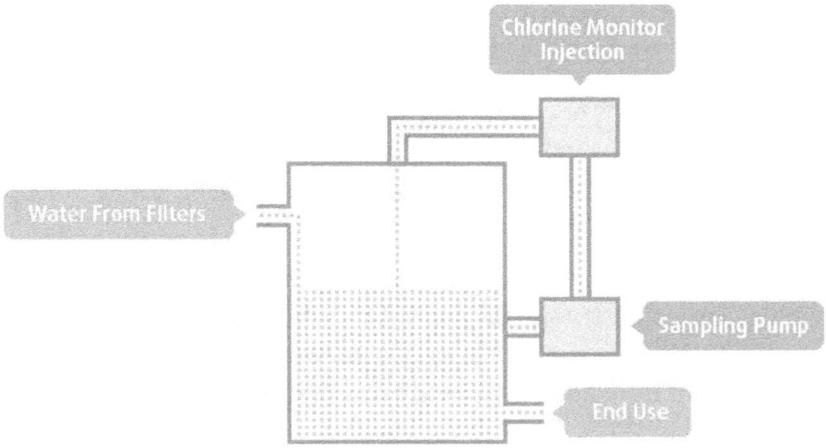

Fig. 4. Advantages of Chlorine Sterilisation.

8.10.4 Ultraviolet Sterilisation

Ultraviolet light is an alternative method of sterilisation. It works by disrupting and damaging pathogens' cells.

UV light sterilisation requires rainwater to be virtually free of any large sediment. If the rainwater has not been filtered, 'shadowing' can occur, whereby sediment blocks the UV light rays, reducing the effectiveness of sterilisation.

Advantages of UV Sterilisation

- Ultraviolet lights bulbs have a lifespan of 12 months to 2 years.
- UV units can handle large flow rates.
- UV light treated water can be used immediately after it finishes treatment.
- Rainwater treated with UV is less of a corrosion risk to exposed metal.

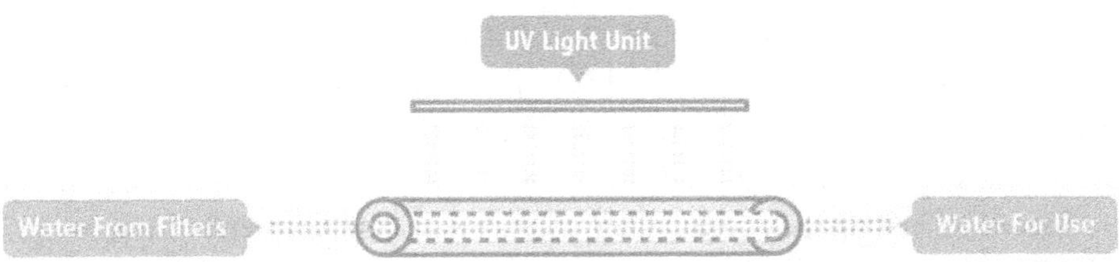

Fig. 5. Rainwater harvesting is an excellent opportunity for businesses aiming to reduce operating costs.

It is often unknown that collected rainwater isn't immediately safe to use until correct rainwater treatment has been carried out. A Cleanawater Rainwater Harvesting system will help you both capture and treat rainwater for your business.

Cleanawater's packaged range of Rainwater Harvesting systems are cost effective, portable and ideally suited for your harvesting application.

8.11 WATER CONSERVATION

There are several ways to reduce the use of water. For example, most irrigation systems waste water; typically, only between 35% and 50% of water withdrawn for irrigated agriculture ever reaches the crops. Most soaks into unlined canals, leaks out of pipes, or evaporates before reaching (or after being applied to) the fields. Swales and cisterns can be used to catch and store excess rainwater.

Water should be used more efficiently in industry, which should use a closed water cycle if possible. Also, industry should prevent polluting water so that it can be returned into the water cycle. Whenever possible, gray wastewater should be used to irrigate trees or lawns. Water drawn from aquifers should be recharged by treating the wastewater and returned to the aquifer.

Water can be conserved by not allowing freshwater to be used to irrigate luxuries such as golf courses. Luxury goods should not be produced in areas where freshwater has been depleted. For example, 1,500 liters of water is used on average for the manufacturing of a single computer and monitor.

8.11.1 Water Management

Sustainable water management involves the scientific planning, developing, distribution and optimization of water resources under defined water polices and regulations. Examples of policies that improve water management include the use of technology for efficiency monitoring and use of water, innovative water prices and markets, irrigation efficiency techniques, and much more.

Experience shows that higher water prices lead to improvements in the efficiency of use—a classical argument in economics, pricing, and markets. For example, Clark County, Nevada, raised its water rates in 2008 to encourage conservation. Economists propose to encourage conservation by adopting a system of progressive pricing whereby the price per unit of water used would start out very small, and then rise substantially for each additional unit of water used. This tiered-rate approach has been used for many years in many places, and is becoming more widespread. A Freakonomics column in the New York Times similarly suggested that people would respond to higher water prices by using less of it, just as they respond to higher gasoline prices by using less of it. The Christian Science Monitor has also reported on arguments that higher water prices curb waste and consumption.

Conversely, certain subsidies can lead to inefficient use of water. Water subsidies often involve contentious policy issues that are political in nature. In 2004, the Environmental Working Group criticized the U.S. government for selling subsidized water to corporate farms for an average price of only $17 per acre foot (326,000 gallons).

In his book The Ultimate Resource 2, Julian Simon claimed that there is a strong correlation between government corruption and lack of sufficient supplies of safe, clean water. Simon wrote, "there is complete agreement among water economists that all it takes to ensure an adequate supply for agriculture as well as for households in rich countries is that there be a rational structure of water law and market pricing. The problem is not too many people but rather defective laws and bureaucratic interventions; freeing up markets in water would eliminate just about all water problems forever... In poor water-short countries the problem with water supply—as with so many other matters—is lack of wealth to create systems to supply water efficiently enough. As these countries become richer, their water problems will become less difficult". This theoretical argument, however, ignores real-world conditions, including strong barriers to open water markets, the difficulty of moving water from one region to another, water rights laws that prevent redistribution based solely on economic value, inability of some populations to pay for water, and grossly imperfect information on water use. Actual experience with peak water constraints in some wealthy, but water-short countries and regions still suggests serious difficulties in reducing water challenges.

8.12 CLIMATE CHANGE

Extensive research has shown the direct links between water resources, the hydrologic cycle, and climatic change. As climate changes, there will be substantial impacts on water demands, precipitation patterns, storm frequency and intensity, snowfall and snowmelt dynamics, and more. Evidence from the IPCC to Working Group II, has shown climate change is already having a direct effect on animals, plants and water resources and systems. A 2007 report by the Intergovernmental Panel on Climate Change counted 75 million to 250 million people across Africa who could face water shortages by 2020. Crop yields could increase by 20% in East and Southeast Asia, but decrease by up to 30% in Central and South Asia. Agriculture fed by rainfall could drop by 50% in some African countries by 2020. A wide range of other impacts could affect peak water constraints.

Loss of biodiversity can be attributed largely to the appropriation of land for agroforestry and the effects of climate change. The 2008 IUCN Red List warns that long-term droughts and extreme weather puts additional stress on key habitats and, for example, lists 1,226 bird species as threatened with extinction, which is one-in-eight of all bird species.

8.12.1 Backstop Water Sources

The concept of a "backstop" resource is a resource that is sufficiently abundant and sustainable to replace non-renewable resources. Thus, solar and other renewable energy sources are considered "backstop" energy options for unsustainable fossil fuels. Similarly, Gleick and Palaniappan defined "backstop water sources" to be those resources that can replace unsustainable and non-renewable use of water, albeit typically at a higher cost. The classic backstop water source is desalination of seawater. If the rate of water production is not sufficient in one area, another "backstop" could be increased interbasin transfers, such as pipelines to carry freshwater from where it is abundant to an area where water is needed. Water can be imported into an area using water trucks.The most expensive and last resort measures of getting water to a community such as desalination, water transfers are called "backstop" water sources. Fog catchers are the most extreme of backstop methods.

To produce that fresh water, it can be obtained from ocean water through desalination. A January 17, 2008 article in the Wall Street Journal stated, "World-wide, 13,080 desalination plants produce more than 12 billion US gallons (45,000,000 m) of water a day, according to the International Desalination Association".Israel is now desalinizing water at a cost of US$0.53 per cubic meter.Singapore is desalinizing water for US$0.49 per cubic meter. After being desalinized at Jubail, Saudi Arabia, water is pumped 200 miles (320 km) inland though a pipeline to the capital city of Riyadh.

However, several factors prevent desalination from being a panacea for water shortages:

- High capital costs to build the desalination plant
- High cost of the water produced
- Energy required to desalinate the water
- Environmental issues with the disposal of the resulting brine
- High cost of transporting water

Nevertheless, some countries like Spain are increasingly relying on desalination because of the continuing decreasing costs of the technology.

At last resort, it is possible in some particular regions to harvest water from fog using nets. The water from the nets drips into a tube. The tubes from several nets lead to a holding tank. Using this method, small communities on the edge of deserts can get water for drinking, gardening, showering and clothes washing. Critics say that fog catchers work in theory but have not succeeded as well in practice. This is due to the high expense of the nets and pipe, high maintenance costs and low quality of water.

An alternative approach is that of the Seawater Greenhouse, which consists in desalinating seawater through evaporation and condensation inside a greenhouse solely using solar energy. Successful pilots have been conducted growing crops in desert locations.

8.13 MOISTURE CONSERVATION

During drought periods or normal dry season, conditions in the dry lowlands, preventing moisture loss from the soil is important. Drought causes water stress and wilting of plant leaves. Low humidity and large temperature changes also affect crop reproduction cycles preventing the formation of seeds and fruit. During drought, soil is exposed to heat and wind and can become very hot during the day and may be blown away as dust. Lack of moisture also causes it to harden and crack enabling pests such as weevils to attack crop roots.

The simplest method for minimising the impacts of high temperature, low humidity and lack of rainfall on cropping areas is to apply mulch. Mulch is a layer of dead vegetation laid on the ground both to control weeds and to minimise moisture loss by evaporation.

Mulch can be spread on gardens after they have been dug and either before or after the crop has been planted. Some crops may be planted through the mulch, whereas for others, it is best to spread the mulch on the soil surface after the crop has started growing.

Mulches shade the soil, lowering surface temperatures and reducing the rate of water loss. They also protect soil from wind erosion and frost. Furthermore, in the wet-up periods following drought, they can reduce soil erosion by sheltering the soil from heavy rain and preventing excess rain water running off fields too quickly.

Cultivation under tree canopies is another option for conserving moisture in cropping systems. The trees shade the crops and the soil, lowering soil temperatures and thus minimising evaporation losses.

In the present scenario management and distribution of water has become centralized. People depend on government system, which has resulted in disruption of community participation in water management and collapse of traditional water harvesting system. As the water crisis continues to become severe, there is a dire need of reform in water management system and revival of traditional systems. Scientific and technological studies needs to be carried out to assess present status so as to suggest suitable mitigative measures for the revival to traditional system/wisdom. Revival process should necessarily be backed by people's initiative and active public participation.

Living creatures of the universe are made of five basic elements, viz., Earth, Water, Fire, Air and Sky, Obviously, water is one of the most important elements and no creature can survive without it. Despite having a great regard for water, we seem to have failed to address this sector seriously. Human being could not save and conserve water and it sources, probably because of its availability in abundance. But this irresponsible attitude resulted in deterioration of water bodies with respect to quantity and quality both. Now, situation has arrived when even a single drop of water matters. However. " better late than never", we have not realized the seriousness of this issue and initiated efforts to overcome those problems.

System of collection rainwater and conserving for future needs has traditionally been practiced in India. The traditional systems were time-tested wisdom of not only appropriate technology of Rainwater Harvesting, but also water management systems, where conservation of water was the prime concern. Traditional water harvesting systems were Bawaries, step wells, jhiries, lakes, tanks etc. These were the water storage bodies to domestic and irrigation demands. People were themselves responsible for maintenance to water sources and optimal use of water that could fulfill their needs.

8.14 COMPONENTS OF THE ROOF TOP RAINWATER HARVESTING SYSTEM

The illustrative design of the basic components of roof top rainwater harvesting system is given in the following typical schematic diagram. The system mainly constitutes of following sub components:

- Catchment
- Transportation

The surface that receives rainfall directly is the catchment of rainwater harvesting system. It may be terrace, courtyard, or paved or unpaved open ground. The terrace may be flat RCC/stone roof or sloping roof. Therefore the catchment is the area, which actually contributes rainwater to the harvesting system.

8.14.1 Transportation

Rainwater from rooftop should be carried through down take water pipes or drains to storage/harvesting system. Water pipes should be UV resistant (ISI HDPE/PVC pipes) of required capacity. Water from sloping roofs could be caught through gutters and down take pipe. At terraces, mouth of the each drain should have wire mesh to restrict floating material.

First flush is a device used to flush off the water received in first shower. The first shower of rains needs to be flushed-off to avoid contaminating storable/rechargeable water by the probable contaminants of the atmosphere and the catchment roof. It will also help in cleaning of silt and other material deposited on roof during dry seasons Provisions of first rain separator should be made at outlet of each drainpipe.

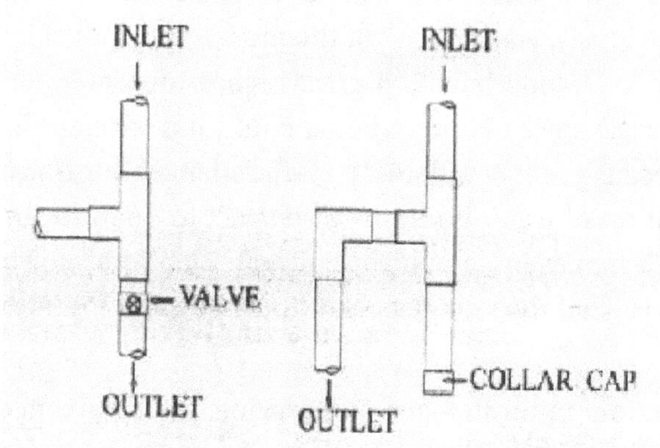

Fig. 6. Transportation.

8.15 FILTER

There is always some skepticism regarding Roof Top Rainwater Harvesting since doubts are raised that rainwater may contaminate groundwater. There is remote possibility of this fear coming true if proper filter mechanism is not adopted. Secondly all care must be taken to see that underground sewer drains are not punctured and no leakage is taking place in close vicinity. Filters are used fro treatment of water to effectively remove turbidity, colour and microorganisms. After first flushing of rainfall, water should pass through filters. There are different types of filters in practice, but basic function is to purify water.

8.15.1 Sand Gravel Filter

These are commonly used filters, constructed by brick masonry and filleted by pebbles, gravel, and sand as shown in the figure. Each layer should be separated by wire mesh.

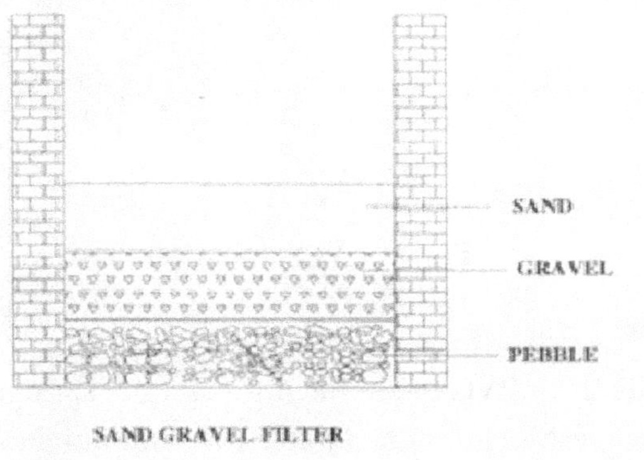

Fig. 7. **Sand gravel filter.**

8.15.2 Charcoal Filter

Charcoal filter can be made in-situ or in a drum. Pebbles, gravel, sand and charcoal as shown in the figure should fill the drum or chamber. Each layer should be separated by wire mesh. Thin layer of charcoal is used to absorb odor if any.

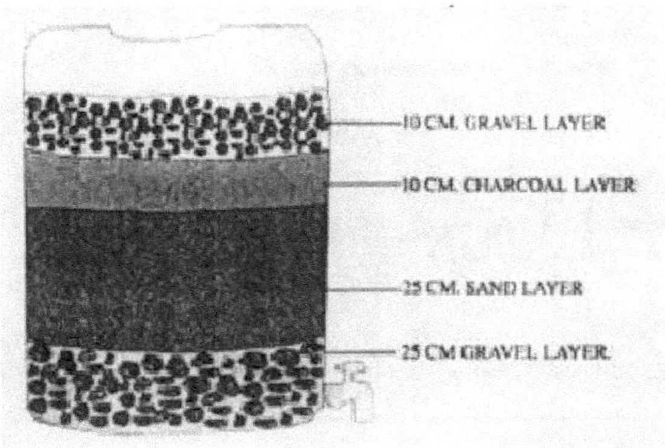

Fig. 8. Charcoal filter.

8.15.3 PVC-Pipe Filter

This filter can be made by PVC pipe of 1 to 1. 20 m length; Diameter of pipe depends on the area of roof. Six inches dia. pipe is enough for a 1500 Sq. Ft. roof and 8 inches dia. pipe should be used for roofs more then 1500 Sq. Ft. Pipe is divided into three compartments by wire mesh. Each component should be filled with gravel and sand alternatively as shown in the figure. A layer of charcoal could also be inserted between two layers. Both ends of filter should have reduce of required size to connect inlet and outlet. This filter could be placed horizontally or vertically in the system.

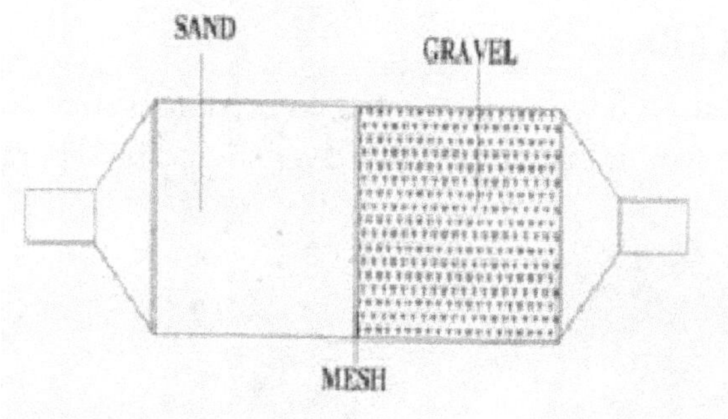

Fig. 9. PVC pipe filter.

8.15.4 Sponge Filter

It is a simple filter made from PVC drum having a layer of sponge in the middle of drum. It is the easiest and cheapest form filter, suitable for residential units.

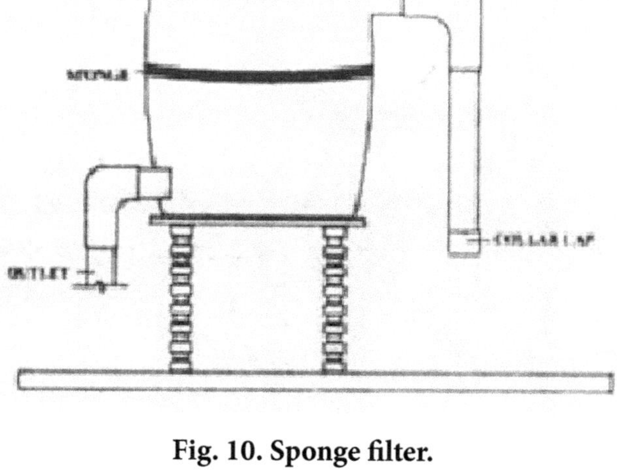

Fig. 10. Sponge filter.

8.16 ADVANTAGES OF RAIN WATER HARVESTING

Rainwater harvesting provides an independent water supply during regional water restrictions and in developed countries is often used to supplement the main supply. It provides water when there is a drought, can help mitigate flooding of low-lying areas, and reduces demand on wells which may enable groundwater levels to be sustained. It also helps in the availability of potable water as rainwater is substantially free of salinity and other salts. Application of rainwater harvesting in urban water system provides a substantial benefit for both water supply and wastewater subsystems by reducing the need for clean water in water distribution system, less generated stormwater in sewer system, as well as a reduction in stormwater runoff polluting freshwater bodies. There has been a large body of work focused on the development of Life Cycle Assessment and Life Cycle Costing methodologies to assess the level of environmental impacts and money that can be saved by implementing rainwater harvesting systems.

More development and knowledge is required to understand the benefits rainwater harvesting can provide to agriculture. Many countries especially those with an arid environment use rainwater harvesting as a cheap and reliable source of clean water. To enhance irrigation in arid environments, ridges of soil are constructed in order to trap and prevent rainwater from running down hills and slopes. Even in periods of low rainfall, enough water is collected in order for crops to grow. Water can be collected from roofs, dams, and ponds can be constructed in order to hold large quantities of rainwater so that even on days where there is little to no rainfall, there is enough available to irrigate crops.

8.17 METHODS OF ROOF TOP RAINWATER HARVESTING

8.17.1 Storage of Direct Use

In this method rain water collected from the roof of the building is diverted to a storage tank. The storage tank has to be designed according to the water requirements, rainfall and catchment availability. Each drainpipe should have mesh filter at mouth and first flush device followed by filtration system before connecting to the storage tank. It is advisable that each tank should have excess water over flow system.

Excess water could be diverted to recharge system. Water from storage tank can be used for secondary purposes such as washing and gardening etc. This is the most cost effective way of rainwater harvesting. The main advantage of collecting and using the rainwater during rainy season is not only to save water from conventional sources, but also to save energy incurred on transportation and distribution of water at the doorstep. This also conserve groundwater, if it is being extracted to meet the demand when rains are on.

8.17.2 Recharging Ground Water Aquifers

Ground water aquifers can be recharged by various kinds of structures to ensure percolation of rainwater in the ground instead of draining away from the surface. Commonly used recharging methods are:-

(a) Recharging of bore wells.

(b) Recharge pits.

(c) Soak ways or Recharge Shafts.

(d) Recharging of dug wells.

(e) Recharge Trenches.

(f) Percolation Tanks

Recharging of Bore Wells

Rainwater collected from rooftop of the building is diverted through drainpipes to settlement or filtration tank. After settlement filtered water is diverted to bore wells to recharge deep aquifers. Abandoned bore wells can also be used for recharge.

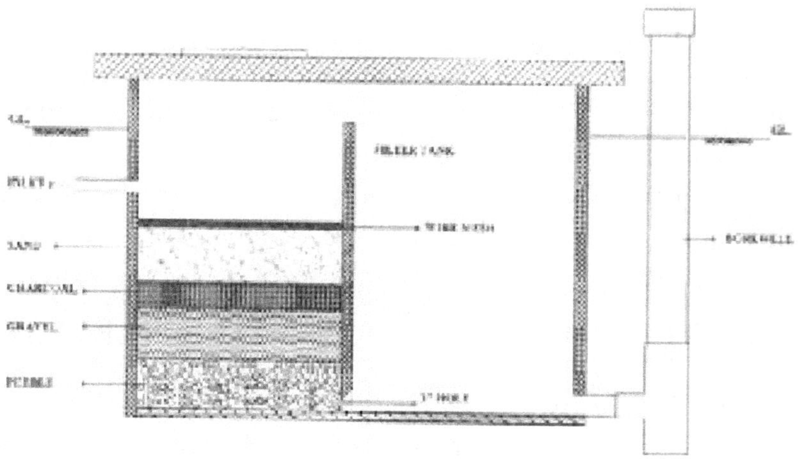

Fig. 11 Filtration tank.

Optimum capacity of settlement tank/filtration tank can be designed on the basis of area of catchement, intensity of rainfall and recharge rate as discussed in design parameters. While recharging, entry of floating matter and silt should be restricted because it may clog the recharge structure. "first one or two shower should be flushed out through rain separator to avoid contamination. This is very important, and all care should be taken to ensure that this has been done. "

Recharge Pits

Recharge pits are small pits of any shape rectangular, square or circular, contracted with brick or stone masonry wall with weep hole at regular intervals. to of pit can be covered with perforated covers. Bottom of pit should be filled with filter media.

The capacity of the pit can be designed on the basis of catchment area, rainfall intensity and recharge rate of soil. Usually the dimensions of the pit may be of 1 to 2 m width and 2 to 3 m deep depending on the depth of pervious strata. These pits are suitable for recharging of shallow aquifers, and small houses.

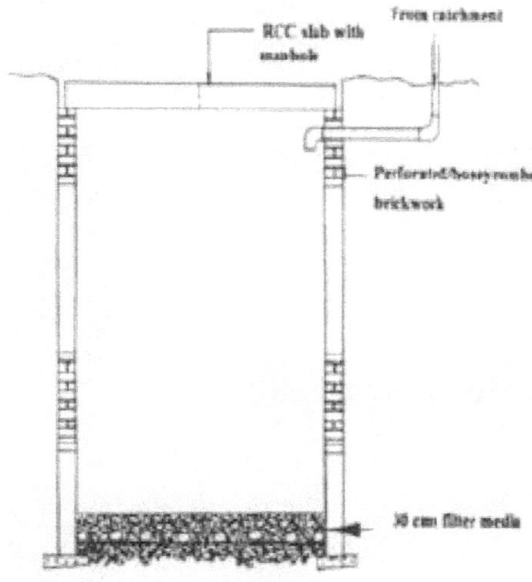

Fig. 12 Typical recharge pit.

Soak Away or Recharge Shafts

Soak away or recharge shafts are provided where upper layer of soil is alluvial or less pervious. These are bored hole of 30 cm dia. up to 10 to 15 m deep, depending on depth of pervious layer. Bore should be lined with slotted/perforated PVC/MS pipe to prevent collapse of the vertical sides. At the top of soak away required size sump is constructed to retain runoff before the filters through soak away. Sump should be filled with filter media.

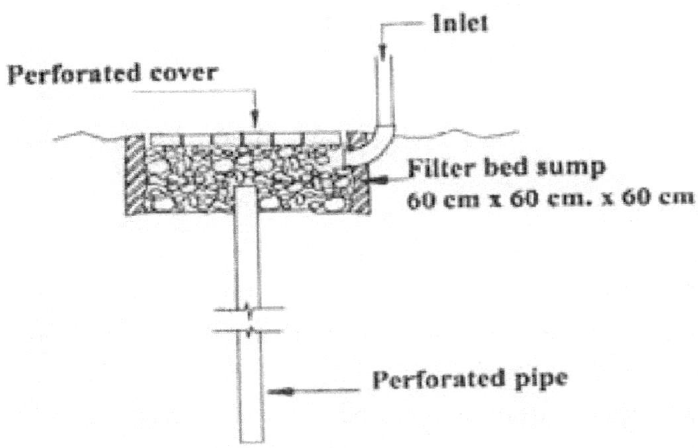

Fig. 13 Soke away or recharge shafts.

Recharging of Dug Wells

Dug well can be used as recharge structure. Rainwater from the rooftop is diverted to dug wells after passing it through filtration bed. Cleaning and desalting of dug well should be done regularly to enhance the recharge rate. The filtration method suggested for bore well recharging could be used.

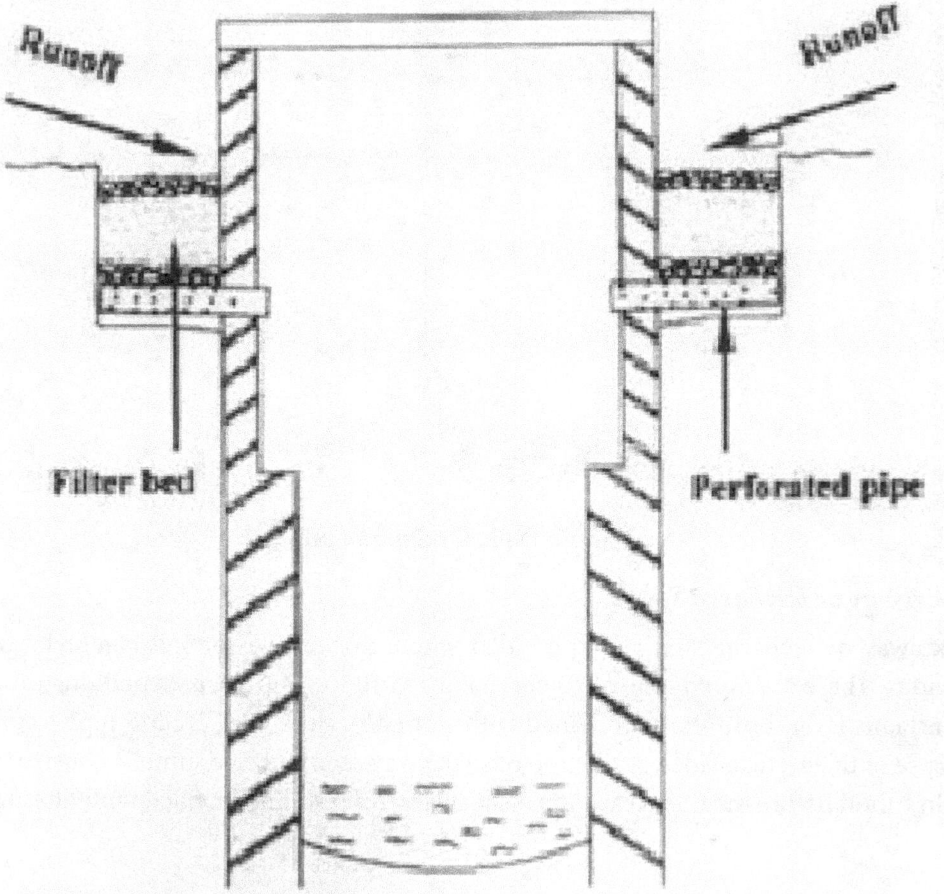

Fig. 14 Recharging of dug welb.

Recharge Trenches

Recharge trench in provided where upper impervious layer of soil is shallow. It is a trench excavated on the ground and refilled with porous media like pebbles, boulder or brickbats. it is usually made for harvesting the surface runoff. Bore wells can also be provided inside the trench as recharge shafts to enhance percolation. The length of the trench is decided as per the amount of runoff expected. This method is suitable for small houses, playgrounds, parks and roadside drains. The recharge trench can be of size 0.50 to 1.0 m wide and 1.0 to 1. 5 m deep.

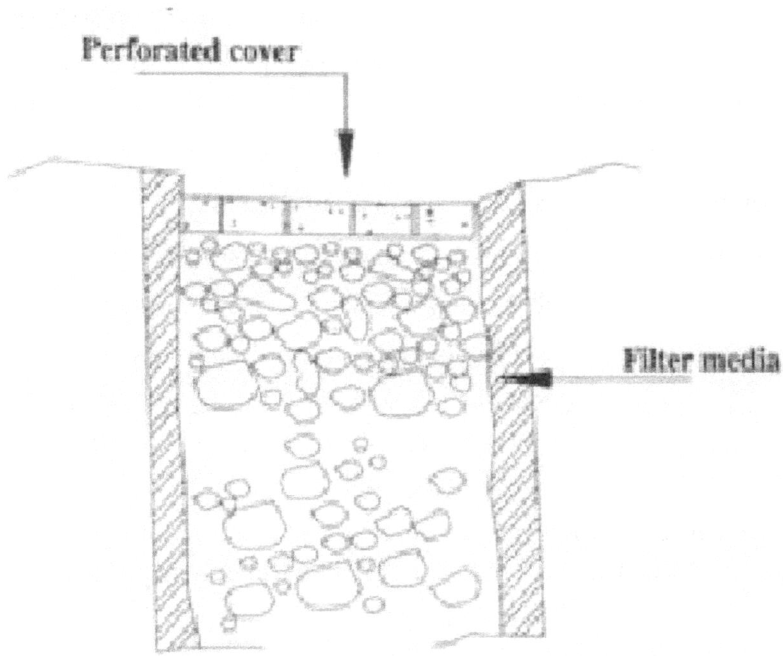

Fig. 15 Rechage Trenches.

Percolation Tanks

Percolation tanks are artificially created surface water bodies, submerging a land area with adequate permeability to facilitate sufficient percolation to recharge the ground water. These can be built in big campuses where land is available and topography is suitable.

Surface run-off and roof top water can be diverted to this tank. Water accumulating in the tank percolates in the solid to augment the ground water. The stored water can be used directly for gardening and raw use. Percolation tanks should be built in gardens, open spaces and roadside green belts of urban area.

Do's and Don'ts

Harvested rainwater is used for direct usage or for recharging aquifers. It is most important to ensure that the rainwater caught is free from pollutants. Following precautionary measures should be taken while harvesting rainwater:-

- Roof or terraces uses for harvesting should be clean, free from dust, algal plants etc.
- Roof should not be painted since most paints contain toxic substances and may peel off.
- Do not store chemicals, rusting iron, manure or detergent on the roof.
- Nesting of birds on the roof should be prevented.

- Terraces should not be used for toilets either by human beings or by pets.
- Provide gratings at mouth of each drainpipe on terraces to trap leaves debris and floating materials.
- Provision of first rain separator should be made to flush off first rains.
- Do not use polluted water to recharge ground water.
- Ground water should only be recharged by rainwater.
- Before recharging, suitable arrangements of filtering should be provided.
- Filter media should be cleaned before every monsoon season.
- During rainy season, the whole system (roof catchment, pipes, screens, first flush, filters, tanks) should be checked before and after each rain and preferably cleaned after every dry period exceeding a month.

At the end of the dry season and just before the first shower of rain is anticipated, the storage tank should be scrubbed and flushed off all sediments and debris.

9

MANAGEMENT OF WASTE WATER

9.1. INTRODUCTION

Wastewater is and will always be with us because we cannot survive without water. When water supplied is used for the numerous human activities, it becomes contaminated or its characteristics is changed and therefore become wastewater. Wastewater can and must be treated to ensure a safe environment and foster public health. There are conventional and non-conventional methods of wastewater treatment and the choice of a particular method should be based on factors such as characteristics of wastewater whether it from a municipality or industry (chemical, textile, pharmaceutical etc.), technical expertise for operation and maintenance, cost implications, power requirements among others. In most developing countries like Ghana, low-cost, low-technology methods such as waste stabilization ponds have been successful whilst conventional methods like trickling filters and activated sludge systems have broken down. Effluent which meets set discharge standards can be appropriately used for aquaculture and also irrigation. Though there are a few challenges in waste water management, they can be surmounted if attention and the necessary financial support is given to it.

Wastewater is water whose physical, chemical or biological properties have been changed as a result of the introduction of certain substances which render it unsafe for some purposes such as drinking. The day to day activities of man is mainly water dependent and therefore discharge 'waste' into water. Some of the substances include body wastes (faeces and urine), hair shampoo, hair, food scraps, fat, laundry powder, fabric conditioners, toilet paper, chemicals, detergent, household cleaners, dirt, micro-organisms (germs) which can make people ill and damage the environment. It is known that much of water supplied ends up as wastewater which makes its treatment very important. Wastewater treatment is the process and technology that is used to remove most of the contaminants that are found in wastewater to ensure a sound environment and good public health. Wastewater Management therefore means handling wastewater to protect the environment to ensure public health, economic, social and political soundness.

Wastewater treatment, also called sewage treatment, the removal of impurities from wastewater, or sewage, before it reaches aquifers or natural bodies of water such as rivers, lakes, estuaries, and oceans. Since pure water is not found in nature (i.e., outside chemical laboratories), any distinction between clean water and polluted water depends on the type and concentration of impurities found in the water as well as on its intended use. In broad terms, water is said to be polluted when it contains enough impurities to make it unfit for a particular use, such as drinking, swimming, or fishing. Although water quality is affected by natural conditions, the word pollution usually implies human activity as the source of contamination. Water pollution, therefore, is caused primarily by the drainage of contaminated wastewater into surface water or groundwater, and wastewater treatment is a major element of water pollution control.

9.2 HISTORY OF WASTEWATER TREATMENT

Wastewater treatment is a fairly new practice although drainage systems were built long before the nineteenth century. Before this time, "night soil" was placed in buckets along streets and workers emptied them into "honeywagon" tanks. This was sent to rural areas and disposed off over agricultural lands. In the nineteenth century, flush toilets led to an increase in the volume of waste for these agricultural lands. Due to this transporting challenge, cities began to use drainage and storm sewers to convey wastewater into waterbodies against the recommendation of Edwin Chadwick in 1842 that "rain to the river and sewage to the soil". The discharge of waste into water courses led to gross pollution and health problems for downstream users.

In 1842, an English engineer named Lindley built the first "modern" sewerage system for wastewater carriage in Hamburg, Germany. The improvement of the Lindley system is basically in improved materials and the inclusion of manholes and sewer appurtenances—the Lindley principles are still upheld today. Treatment of wastewater became apparent only after the assimilative capacity of the waterbodies was exceeded and health problems became intolerable. Between the late 1800s and early 1900s, various options were tried until in 1920, the processes we have today were tried. Its design was however empirical until mid-century. Centralized wastewater systems were designed and encouraged. The cost of wastewater treatment is borne by communities discharging into the plant.

Today there have been great advances to make portable water from wastewater. In recent times, regardless of the capacity of the receiving stream, a minimum treatment level is required before discharge permits are granted. Also presently, the focus is shifting from centralized systems to more sustainable decentralized wastewater treatment (DEWATS) especially for developing countries like Ghana where wastewater infrastructure is poor and conventional methods are difficult to manage.

9.3 OBJECTIVES OF 0WASTEWATER TREATMENT

Wastewater treatment is very necessary for the above-mentioned reasons. It is more vital for the:

9.3.1 Reduction of biodegradable organic substances in the environment: organic substances such as carbon, nitrogen, phosphorus, sulphur in organic matter needs to be broken down by oxidation into gases which is either released or remains in solution.

9.3.2 Reduction of nutrient concentration in the environment: nutrients such as nitrogen and phosphorous from wastewater in the environment enrich water bodies or render it eutrophic leading to the growth of algae and other aquatic plants. These plants deplete oxygen in water bodies and this hampers aquatic life.

9.3.3 Elimination of pathogens: organisms that cause disease in plants, animals and humans are called pathogens. They are also known as micro-organisms because they are very small to be seen with the naked eye. Examples of micro-organisms include bacteria (e.g.vibro cholerae), viruses (e.g. enterovirus, hepatits A & E virus), fungi (e.g. candida albicans), protozoa (e.g entamoeba hystolitica, giardia lamblia) and helminthes (e.g. schistosomamansoni, asaris lumbricoides). These micro-organisms are excreted in large quantities in faeces of infected animals and humans.

9.3.4 Recycling and Reuse of water: Water is a scarce and finite resource which is often taken for granted. In the last half of the 20th century, population has increased resulting in pressure on the already scarce water resources. Urbanization has also changed the agrarian nature of many areas. Population increase means more food has to be cultivated for the growing population and agriculture as we know is by far the largest user of available water which means that economic growth is placing new demands on available water supplies. The temporal and spatial distribution of water is also a major challenge with groundwater resources being overdrawn. It is for these reasons that recycling and reuse is crucial for sustainability.

9.4 TYPES OF WASTEWATERS

Wastewater can be described as in the figure below:

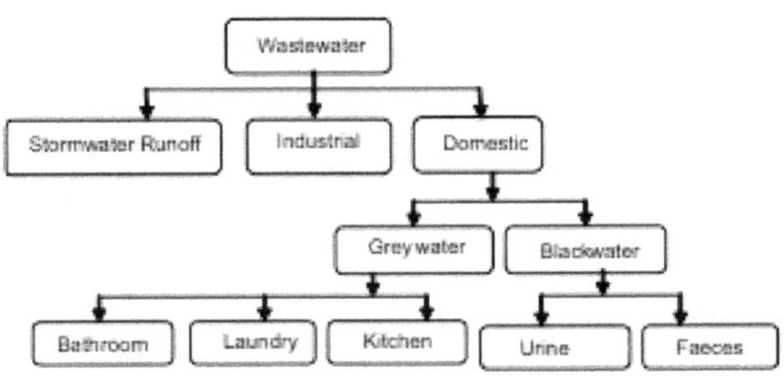

Fig. 1. Types of Wastewater

9.5 CHARACTERISTICS OF WASTEWATER

Depending on its source, wastewater has peculiar characteristics. Industrial wastewater with characteristics of municipal or domestic wastewater can be discharged together. Industrial wastewater may require some pretreatment if it has to be discharged with domestic wastewater. The characteristics of wastewater vary from industry to industry andtherefore would have different treatment processes—for example a cocoa processing company may have a skimming tank in its preliminary treatment stage to handle for instance spilt cocoa butter while a beverage plant may skip this in the design. In general, the contaminants in wastewater are categorized into physical, chemical and biological. Some indicator measured to ascertain these contaminants include.

9.5.1 Physical

- Electrical Conductivity (EC) indicates the salt content
- Total Dissolved Solids (TDS) comprise inorganic salts and small amounts of organic matter dissolved in water
- Suspended solids (SS) comprises solid particles suspended (but not dissolved) in water

9.5.2 Chemical

- Dissolved Oxygen (DO) indicates the amount of oxygen in water
- Biochemical oxygen demand (BOD) indicates the amount of oxygen required by aerobic microorganisms to decompose the organic matter in a sample of water in a defined time period.
- Chemical oxygen demand (COD) indicates the oxygen equivalent of the organic matter content of a sample that is susceptible to oxidation by a strong chemical oxidant
- Total Organic Compound (TOC
- NH4-N and NO3-N show dissolved nitrogen (Ammonium and Nitrate, respectively).
- Total Kjeldhal Nitrogen is a measurement of organically-bound ammonia nitrogen.
- Total-P reflects the amount of all forms of phosphorous in a sample.

9.5.3 Biological

- Total coliforms (TC) is encompassing faecal coliforms as well as common soil microorganisms, and is a broad indicator of possible water contamination.
- Faecal coliforms (FC) is an indicator of water contamination with faecal matter. The common lead indicator is the bacteria Escherichia coli or E. coli.

- Helminth analysis looks for worm eggs in the water

9.6 PROCESS OF WASTEWATER TREATMENT

Due to the nature of contaminants in wastewater—physical, chemical and biological, the unit operations and processes in wastewater treatment can also be categorized as such. The units operations and processes in Waste-water treatment are summarized as follows (Economic and Social Commission for Western Asia.

9.6.1 Physical unit operations

- Screening
- Comminution
- Flow equalization
- Sedimentation
- Flotation
- Granular-medium filtration

9.6.2 Chemical unit operations

- Chemical precipitation
- Adsorption
- Disinfection
- Dechlorination
- Other chemical applications

9.6.3 Biological unit operations

- Activated sludge process
- Aerated lagoon
- Trickling filters
- Rotating biological contactors
- Pond stabilization
- Anaerobic digestion

9.7 LEVELS OF WASTEWATER TREATMENT

There are three broad levels of treatment: primary, secondary and tertiary. Sometimes, preliminary treatment precedes primary treatment.

9.7.1 Preliminary treatment: removes coarse suspended and grits. These can be removed by screening, and grit chambers respectively. This enhances the operation and maintenance of subsequent treatment units. Flow measurement devices, often standing-wave flumes, are necessary at this treatment stage (FAO, 2006).

9.7.2 Primary treatment removes settleable organic and inorganic solids by sedimentation and floating materials (scum) by skimming. Up to 50% of BOD5, 70% of suspended solids and 65% of grease and oil can be removed at this stage. Some organic nitrogen, organic phosphorus, and heavy metals are also removed. Colloidal and dissolved constituents are however not removed at this stage. The effluent from primary sedimentation units is referred to as primary effluent (FAO, 2006).

9.7.3 Secondary treatment is the further treatment of primary effluent to remove residual organics and suspended solids. Also biodegradable dissolved and colloidal organic matter is removed using aerobic biological treatment processes. The removal of organic matter is when nitrogen compounds and phosphorus compounds and pathogenic microorganisms are removed. The treatment can be done mechanically like in trickling filters, activated sludge methods rotating biological contactors (RBC) or non-mechanically like in anaerobic treatment, oxidation ditches, stabilization ponds etc.

9.7.4 Tertiary treatment or advance treatment is employed when specific wastewater constituents which cannot be removed by secondary treatment must be removed. Advance treatment removes significant amounts of nitrogen, phosphorus, heavy metals, biodegradable organics, bacteria and viruses. Two methods can be used effectively to filter secondary effluent—traditional sand (or similar media) filter and the newer membrane materials. Some filters have been improved, and both filters and membranes also remove helminths. The latest method is disk filtration which utilizes large disks of cloth media attached to rotating drums for filtration.

At this stage, disinfection by the injection of Chlorine, Ozone and Ultra Violet (UV) irradiation can be done to make water meet current international standards for agricultural and urban re-use.

9.8 . METHODS OF WASTEWATER TREATMENT

There are conventional and non-conventional wastewater treatment methods which have been proven and found to be efficient in the treatment of wastewater. Conventional methods compared to non-conventional wastewater treatment methods has a relatively highlevel of automation. Usually have pumping and power requirements. They require skilled labour for operation and maintenance of the system.

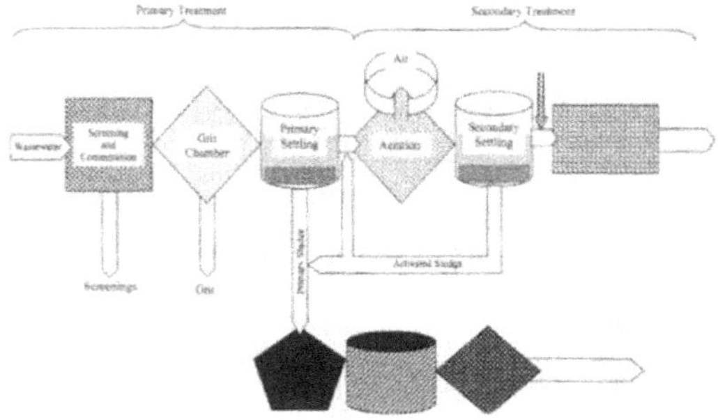

Fig. 2. Typical Wastewater Treatment Plant (Source: NPTEL (accessed 2010))

9.8.1 Conventional methods

Examples of conventional wastewater treatment methods include activated sludge, trickling filter, rotating biological contactor methods. Trickling filters and Rotating Biological Contactors are temperature sensitive, remove less BOD, and trickling filters cost more to build than activated sludge systems. Activated sludge systems are much more expensive to operate because energy is needed to run pumps and blowers (National Programme on Technology Enhanced Learning (NPTEL), 2010).

9.8.1.1 Activated sludge

Activated sludge refers to biological treatment processes that use a suspended growth of organisms to remove BOD and suspended solids. It is based on the principle that intense wastewater aeration to forms flocs of bacteria (activated sludge), which degrade organic matter and be separated by sedimentation. The system consists of aeration and settling tanks with other appurtenances such as return and waste pumps, mixers and blowers for aeration and a flow measurement device. To maintain the concentration of active bacteria in the tank, part of the activated sludge is recycled.

Primary effluent (or plant influent) is mixed with return activated sludge to form mixed liquor which is aerated for a specified length of time. By aerating the system, activated sludge organisms use the available organic matter as food, thereby, producing stable solids and more organisms. The suspended solids produced by the process and the additional organisms become part of the activated sludge. The solids are then separated from thewastewater in the settling tank and are returned to the influent of the aeration tank (return activated sludge). Periodically the excess solids and organisms are removed from the system (waste activated sludge) to enhance the performance of the system.

Factors such as temperature, return rates, amount of oxygen available, amount of organic matter available, pH, waste rates, aeration time, and wastewater toxicity affect the performance of an activated sludge treatment system. A balance therefore must be maintained between the amount of food (organic matter), organisms (activated sludge) and dissolved oxygen (NPTEL, 2010).

Activated Sludge systems are requires less space compared to trickling filter and has high effluent quality. The disadvantage is that BOD is higher at one end of the tank than the other the microorganisms will be physiologically more active at that end than the other unless a complet mixing activated sludge system process is used. Presently there are 11 activated sludge plants in Ghana, mainly installed by the large hotel.

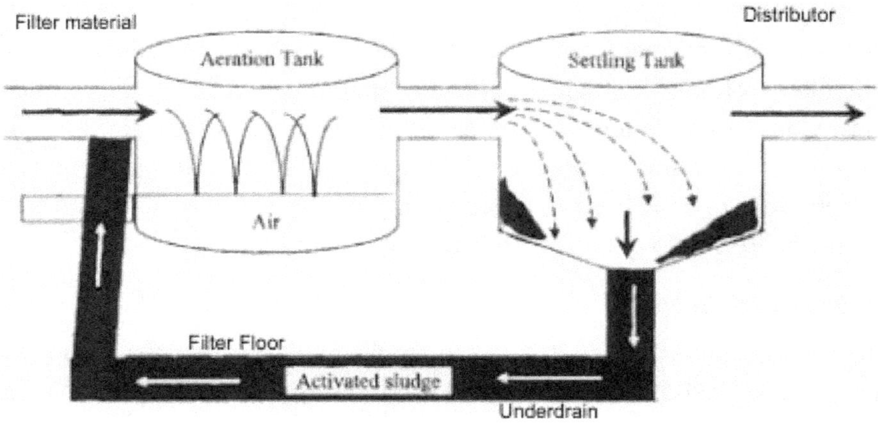

Fig. 3. An activated Sludge System (Source: Mountain Empire College, 2010)

9.8.1.2 Trickling filter

It is a growth process in which microorganisms responsible for treatment are attached to an inert packing material. It is made up of a round tank filled with a carrier material (volcanic rock, gravel or synthetic material). Wastewater is supplied from above and trickles through filter media allowing organic material in the wastewater to be adsorbed by a population of microorganisms (aerobic, anaerobic, and facultative bacteria; fungi; algae; and protozoa) attached to the medium as a biological film or slime layer (approximately 0.1 to 0.2 mm thick).

Degradation of organic material by the aerobic microorganisms in the outer part of the slime layer occurs. As the layer thickens through microbial growth, oxygen cannot penetrate the medium face, and anaerobic organisms develop. The biological film continues to grow to such a point that microorganisms near the surface cannot cling to the medium, and a portion of the slime layer falls off the filter. This process is known as sloughing. The sloughed solids are picked up by the underdrain system and transported to a clarifier for removal from the wastewater (US EPA, 2000).

Trickling filters are efficient in that effluent quality in terms of BOD and suspended solids removal is high. Its operational costs are relatively low due to low electricity requirements. The process is simpler compared to activated sludge process or some package treatment plants. Its operation and maintenance requirements is however high due to the use of electrical power. Skilled labour is required to keep the trickling filter running trouble-free: e.g. prevent clogging, ensure adequate flushing, control filter flies. It is suitable for some relatively wealthy, densely populated areas which have a sewerage system and centralized wastewater treatment; also suitable for greywater treatment.

It also requires more space compared to some other technologies and has potential for odour and filter flies (NPTEL, 2010). This method has been widely used in Ghana. There are 14 trickling filter plants in Accra though they have broken down.

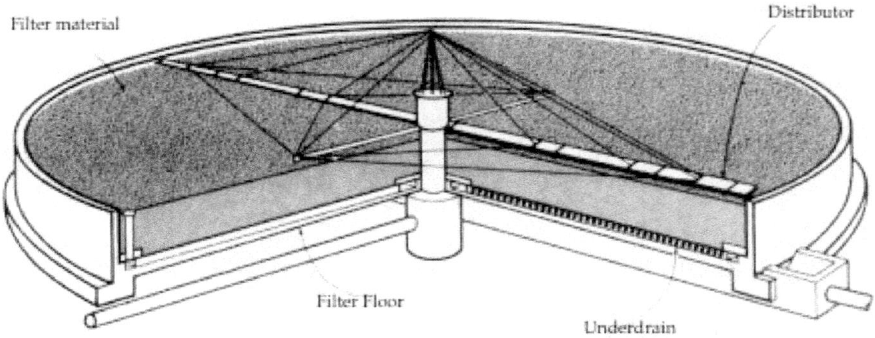

Fig. 4. Cross section of a trickling filter (Source: ESCWA, 2003)

9.8.1.3 Rotating biological contactors

Rotating biological contactors (RBCs) consist of vertically arranged, plastic media on a horizontal, rotating shaft. The plastics range from 2 – 4 m in diameter and up to 10 mm thick. The biomass-coated media are alternately exposed to wastewater and atmospheric oxygen as the shaft slowly rotates at 1–1.5 rpm (necessary to provide hydraulic shear for sloughing and to maintain turbulence to keep solid in suspension), with about 40% of the media submerged. High surface area allows a large, stable biomass population to develop, with excess growth continuously and automatically shed and removed in a downstream clarifier. Thichness of biofilm may reach 2 – 4 mm depending on the strength of wastewater and the rotational speed of the disk. RBC systems are relatively new, though it appeared to be best suited to treat municipal wastewater, they have been installed in many petroleum facilities because of their ability to quickly recover from upset conditions. The RBC system is easily expandable should the need arise, and RBCs are also very easy to enclose should volatile organic content containment become necessary. RBCs have relatively low power requirements and can even be powered by compressed air which can also aerate the system. They follow simple operating procedures and thus require a moderately skilled labour. RBCs are however capital intensive to install and sensitive to temperature.

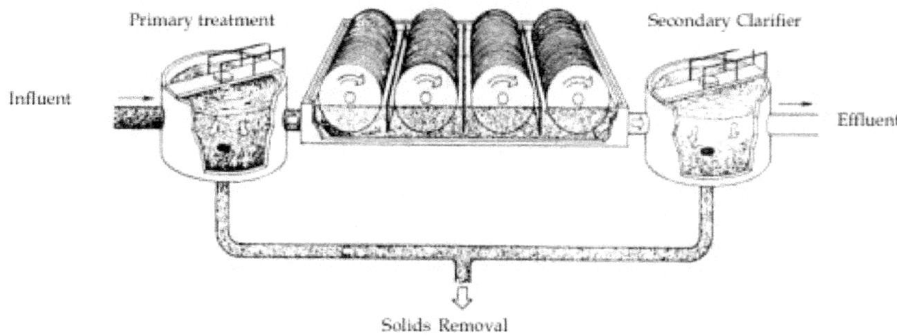

Fig. 5. Rotating Biological Contactors (Source: ESCWA, 2003)

9.8.1.4 Membrane bioreactors

This method performs more than just one treatment step. Membrane bioreactor (MBR) systems are unique processes, which combine anoxic- and aerobic-biological treatment with an integrated membrane system that can be used with most suspended-growth, biological wastewater-treatment systems.

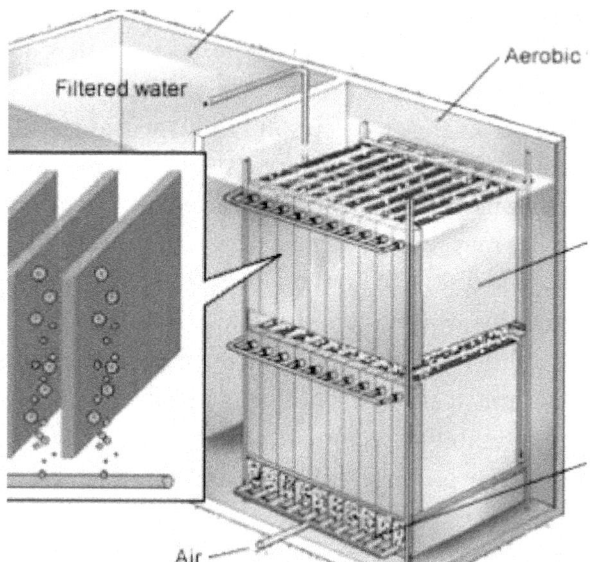

Fig. 6. Membrane Bioreato (Source: Google Images)

Wastewater is screened before entering the biological treatment tank. Aeration within the aerobic-reactor zone provides oxygen for biological respiration and maintains solids in suspension. MBR relies on submerged membranes to retain active biomass in the process. This allows the biological process to operate at longer than normal sludge ages (typically 20- 100 days for a MBR) and to increase mixed-liquor, suspended-solids (MLSS) concentrations (typically 8,000-15,000 mg/l) for more effective removal of pollutants. High MLSS concentrations reduce biological-volume requirements and the associated space needed to only 20–30% of conventional biological processes. MBRs cover a small land area as it eliminates the need for secondary clarifiers, which equates to a huge savings in both footprint and concrete costs. They can operate at higher biomass concentrations (MLSS) than conventional treatment processes. Facility can be expanded by simply adding more membranes to existing basins without expanding land cover. For reuse quality, it does not require tertiary treatment, polymer addition, or any further treatment processes to meet standards. This reduction in the number of unit processes further improves system reliability and reduces operation activities. The generally high effluent quality reduces the burden on disinfection in the treatment process.

9.9.NON-CONVENTIONAL METHODS

These are low-cost, low-technology, less sophisticated in operation and maintenance biological treatment systems for municipal wastewater. Although these systems are land intensive by comparison with the conventional high-rate biological processes, they are often more effective in removing pathogens and do so reliably and continuously if system is properly designed and not overloaded. Some of the non-conventional methods include stabilization ponds, constructed wetlands, oxidation ditch, soil aquifer treatment.

9.9.1 Waste stabilization ponds

Waste Stabilization Ponds are man-made, shallow basins which comprise of a single series or several series of anaerobic, facultative or maturation ponds. This is a low-technology treatment process with 4 or 5 ponds of different depths with different biological activities. Treatment of the wastewater occurs as constituents are removed by sedimentation or transformed by biological and chemical processes.The anaerobic ponds are mainly designed for the settling and removal of suspended solids as well as the breakdown of some organic matter (BOD_5). In facultative ponds, organic matter is further broken down to carbon dioxide, nitrogen and phosphorous by using oxygen produced by algae in the pond. Maturation ponds usually remove nutrients and pathogenic microorganisms, thus primary treatment occurs in anaerobic ponds while secondary and tertiary treatment occurs in facultative and maturation ponds respectively (Awuah, 2002). Anaerobic ponds are usually between 2-5 m deep and receive high organic loads equivalent to 100g BOD5 and m^3/d leading to anaerobic conditions throughout the pond. If properly designed, anaerobic ponds can remove 60% of BOD5 at 20^0 C. Facultative ponds are 1-2 m deep and usually receive the effluent from an anaerobic pond. In some designs, they receive raw wastewater acting as primary facultative pond. In facultative ponds organic loads are lower and allows for algal growth which accounts for the dark green colour of wastewater. Algae and aerobic bacteria generate oxygen which breaks down BOD5. Good wind velocity generates mixing of wastewater in ponds thus leading to uniform mixing of BOD5, oxygen, bacteria and algae which better stabilizes waste.

Maturation ponds are usually shallow ponds of about 1.0-1.5 m deep allowing aerobic conditions in for the treatment of facultative pond effluents. Further reduction of organic matter, nutrients and pathogenic microorganisms occurs here. Algal population in maturation ponds is more diverse and removal of nitrogen and ammonia is more prominent.

In Ghana so far, stabilization ponds have worked very well due to the convenient climatic conditions. It usually flows under gravity from one pond to the other and mostly does not require any pumping. It is less energy dependent thus plant activities cannot be interrupted due to power cuts. Its disadvantages however include odour problems and it requires a large area of land to function properly. Presently there are 21 stabilisation ponds in Ghana mainly in Accra and Kumasi. Some of them like the Tema Community 3, Achimota, have been closed Various combinations and arrangement of ponds are possible. The figure below shows some possible combinations.

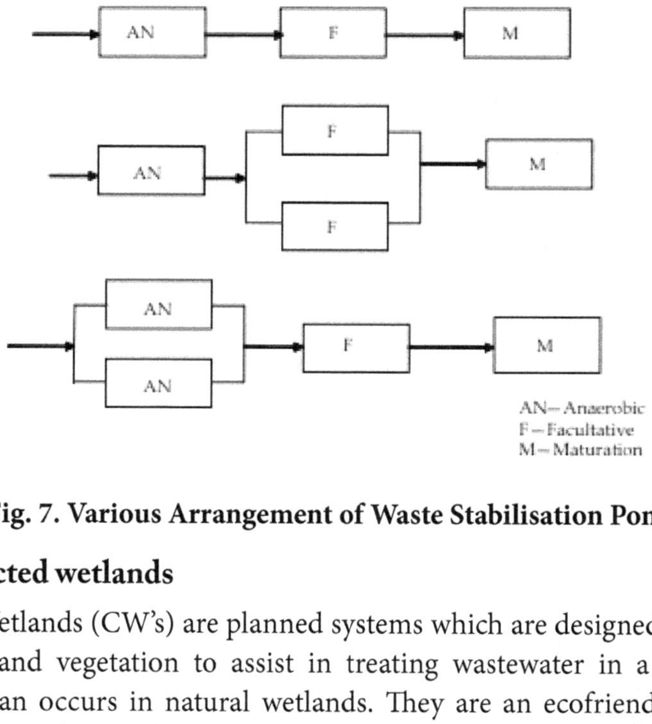

Fig. 7. Various Arrangement of Waste Stabilisation Ponds

9.9.2 Constructed wetlands

Constructed Wetlands (CW's) are planned systems which are designed and constructed to employ wetland vegetation to assist in treating wastewater in a more controlled environment than occurs in natural wetlands. They are an ecofriendly and a suitable alternative for secondary and tertiary treatment of municipal and industrial wastewater. They are suitable for the removal of organic materials, suspended solids, nutrients, pathogens, heavy metals and toxic pollutants. They are not ideal for the treatment of raw sewage, pre-treatment of industrial wastewater to maintain the biological balance of the wetland ecosystem.

There are two types of CW's namely Free Water Surface (FWS) and Subsurface Flow (SSF) systems. As the name suggests, with FWS, water flows above the ground and plants are rooted in the sediment layer below the water column. With SSF, water flows through a porous media such as gravels in which the plants are rooted. From a public health perspective, SSF should be used in primary treatment of wastewater because there is no direct contact of wastewater with atmosphere.

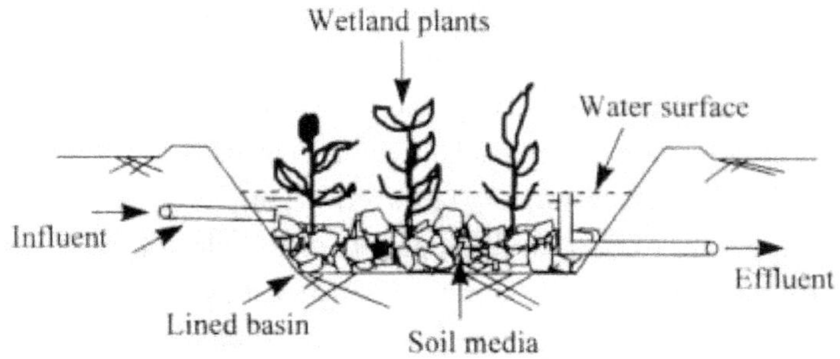

Fig. 8. Free Water Surface System (Source: ESCWA, 2003)

The SSF is mostly anoxic or anaerobic as oxygen supplied by the roots of plants is used up in biofilm growth and as such does not reach the water colomn. The flow of water in SSF can be horizontal or vertical (Kayombo et al., 2000). FWS are suitable for treating secondary and tertiary effluents and also providing habitat due to aerobic conditions at and near the surface of the water. There condition at the bottom sediment is however anoxic. Wetlands plants or macrophytes utilized in CW's include Cattails (Typha latifolia sp), Scirpus (Bulrus), Lemna (duckweed), Eichornia crassipes (water hyacinth), Pistia stratiotes (water lettuce) Hydrocotyle spp. (pennywort), Phragmites (reed) have been known and used in constructed wetlands.

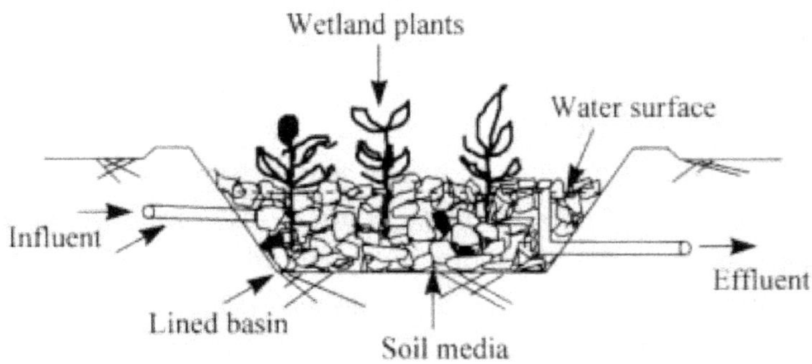

Fig. 9. Sub-surface flow system (Source: ESCWA, 2003)

CW's are relatively cheaper to construct operate and easy to maintain. This is an important decision variable for developing countries. In Egypt, according to Hendy (2006), between 2000 and 2004, a 60 acre artificial wetland constructed cost 25% the cost of conventional sewage treatment plant. They provide effective and reliable treatment of wastewater and are tolerant to fluctuating hydrologic and contaminant loading rates. With the example in Egypt, \$9 million (US) was spent to treat an initial volume of 25,000 metric tons per day. After a year of use, it was determined that the wetland was capable of treating 40,000 metric tons per day. Also a study revealed over 60% removal of BOD5,

COD, nitrogen among others. CW's also provide indirect benefits such as enjoying the scenic views of green spaces, encouraging wildlife habitats and providing recreational and educational centres. Again, in Egypt, the fishing industry is expanding since the wastewater was no longer being discharged directly into the waterways, the local fisheries improved. As reported nitrates and heavy metals were filtered out, leaving the fish healthier, larger and in abundant quantity. This indirectly led to poverty reduction. They however have some disadvantages such as land requirements, its design and operation criteria is presently imprecise. CW's are biologically and hydrologically complex and its process dynamics are not completely understood. Sometimes there are cost implications of gravels fills and site grading during construction. It must be emphasized that if properly designed, constructed wetlands should not breed pests and mosquitoes. In Ghana, there are not many CW's. There is presently a pilot SSF donor CW in Tema. This plant is not entirely low-cost as it was designed with some energy dependent units.

9.9.3 Oxidation ditches

An oxidation ditch is a modified activated sludge biological treatment process that utilizes hydraulic retention time of 24 - 48 hours, and a sludge age of 12 - 20 days. to remove biodegradable organics. Oxidation ditches are typically complete mix systems, but can be modified. Typical oxidation ditch treatment systems consist of a single or multichannel configuration within a ring, or oval. Preliminary treatment, such as bar screens and grit removal, normally precedes the oxidation ditch. Primary settling prior to an oxidation ditch is sometimes practiced and tertiary filters may be required after clarification, depending on the effluent requirements. Disinfection is required and reaeration may be necessary prior to final discharge. Horizontally or vertically mounted aerators provide circulation, oxygen transfer, and aeration in the ditch. Flow to the oxidation ditch is aerated and mixed with return sludge from a secondary clarifier. The mixing process entrains oxygen into the mixed liquor to foster microbial growth and the motive velocity ensures contact of microorganisms with the influent. Aeration increases dissolved oxygen concentration but decreases as biomass takes up oxygen during mixing in the ditch. Solids also remain in suspension during circulation. They require more power than waste stabilization ponds less land, and are easier to control than processes such as activated sludge process. A typical process flow diagram of treatment plant using an oxidation ditch is shown in Figure 10.

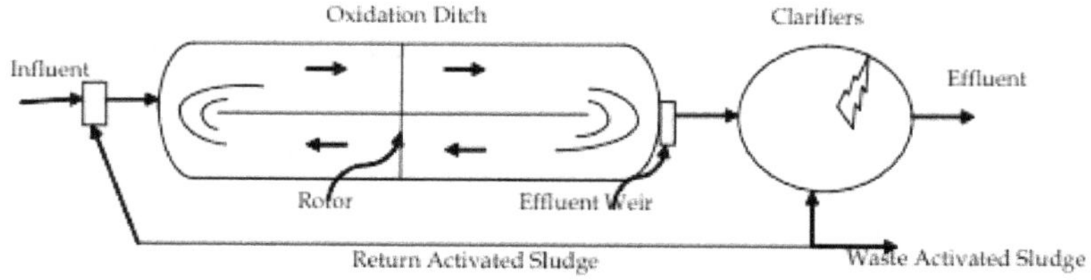

Fig. 10. Oxidation Ditch

9.9.4 Upflow anaerobic sludge blanket (UASB)

Upflow anaerobic sludge blanket is an anaerobic process using blanket of bacteria to absorb polluting load. It is a form of anaerobic digester which forms a blanket of granular sludge which suspends in the tank. Wastewater flows upwards through the blanket and is processed (degraded) by the anaerobic microorganisms. The upward flow combined with the settling action of gravity suspends the blanket with the aid of flocculants. Small sludge granules begin to form whose surface area is covered in agregations of bacteria. In the absence of any support matrix, the flow conditions create a selective environment in which only those microorganisms, capable of attaching to each other, survive and proliferate.

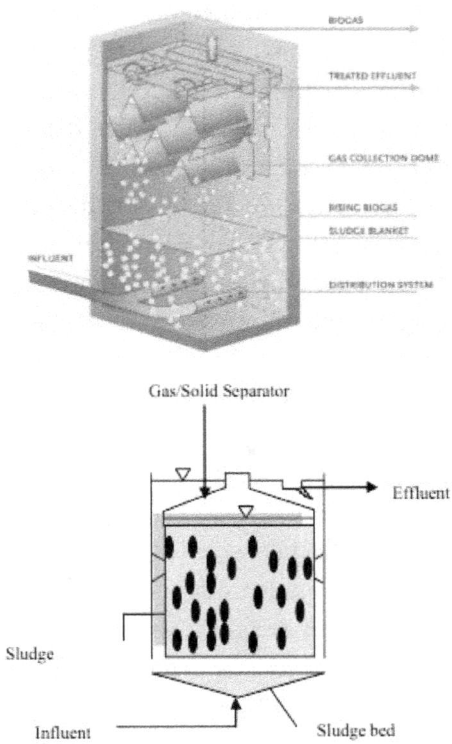

Fig. 11. Upflow Anaerobic Sludge Blanket (Source: Google Images)

Eventually the aggregates form into dense compact biofilms referred to as granules. The UASB reactor works best when desirable micro-organisms are retained as highly active and fast settling granules. In the UASB reactor, when high solids retention time is met, separation of gas, sludge solids from the liquid occurs. The special Gas-Solid-Liquid Separators in the reactor enable collection of biogas and recycle of anaerobic biomass. Biogas contains 50 to 80% methane. UASB is suitable for the primary treatment of high-COD mainly soluble industrial effluents. It can also be used for the treatment of wastewater effluents of low and medium strength. It is suited to hot climates Low energy requirement, less operation and maintenance, lower skill requirement for operation, less sludge production, resource recovery through biogas generation and stabilized waste as manure. UASBs however have relatively poor effluent quality than processes such as activated sludge process. The technology however, needs constant monitoring to ensure that the sludge blanket is maintained, and not washed out. The heat produced as a by-product of electricity generation can be reused to heat the digestion tanks.

9.9.5 Soil quifer treatment

Soil matrix has quite a high capacity for treatment of normal domestic sewage, as long as capacity is not exceeded. Partially-treated sewage effluent is allowed to infiltrate in controlled conditions to the soil. The unsaturated or "vadose" zone then acts as a natural filter and can remove essentially all suspended solids, biodegradable materials, bacteria, viruses, and other microorganisms. Significant reductions in nitrogen, phosphorus, and heavy metals concentrations can also be achieved. After the sewage, treated in passage through the vadose zone, has reached the groundwater it is usually allowed to flow some distance through the aquifer for further purification before it is collected through the aquifer. Soil-aquifer treatment is a low-technology, advanced wastewater treatment system. It also has an aesthetic advantage over conventionally treated sewage since effluent from an SAT system is clear and odour-free and it is viewed as groundwater either than effluent. Discharge effluent should travel sufficient distance through the system and residence times should be long enough, to produce effluent of desired quality.

9.10 FAECAL SLUDGE TREATMENT AND DISPOSAL

Sewage sludge contains organic and inorganic solids that were found in the raw wastewater. Sludge from primary and secondary clarifier as well as from secondary biological treatment need to be treated. The generated sludge is usually in the form of a liquid or semisolid, containing 0.25 to 12 per cent solids by weight, depending on the treatment operations and processes used. Sludge is treated by means of a variety of processes that can be used in various combinations. Thickening, conditioning, dewatering and drying are primarily used to remove moisture from sludge, while digestion, composting, incineration, wet-air oxidation and vertical tube reactors are used to treat or stabilize the organic material in the sludge .

9.10.1 Thickening

Thickening is done to increase the solids content of sludge by the reduction of the liquid content. An increase in solids content from 3 to about 6 per cent can decrease total sludge volume significantly by 50 per cent. Sludge thickening methods are usually physical in nature: they include gravity settling, flotation, centrifugation and gravity belts.

9.10.2 Stabilization

Sludge stabilization is aimed at reducing the pathogen content, eliminate offensive odours, and reduce or eliminate the potential for putrefaction. Some methods used for sludge stabilization include lime stabilization, heat treatment, anaerobic digestion, aerobic digestion and composting.

9.11 WASTEWATER REUSE IN AGRICULTURE

Irrigation with wastewater is both disposal and utilization and indeed is an effective form of wastewater disposal (as in slow-rate land treatment). However, some degree of treatment must normally be provided to raw municipal wastewater before it can be used for agricultural or landscape irrigation or for aquaculture. In many industrialized countries, primary treatment is the minimum level of preapplication treatment required for wastewater irrigation. It may be considered sufficient treatment if the wastewater is used to irrigate crops that are not consumed by humans or to irrigate orchards, vineyards, and some processed food crops . Nutrients in municipal wastewater and treated effluents are a particular advantage as supplemental fertilizers. Success in using treated wastewater for crop production will largely depend on adopting appropriate strategies aimed at optimizing crop yields and quality, maintaining soil productivity and safeguarding the environment. Several alternatives are available and a combination of these alternatives will offer an optimum solution for a given set of conditions. The user should have prior information on effluent supply and its quality. Wastewater effluent can be blended with conventional water or solely used. Heavy metal concentrations in streams used for irrigation in and around urban centres such as Accra and Kumasi have been sometimes found to be beyond recommended levels for irrigation purposed and should therefore may pose a health concern. Countries must develop standards in congruence with the WHO guidelines and enforce it.

9.12 INDUSTRIAL WASTEWATER TREATMENT

In general, the type of plant to be installed depends on the characteristics of the wastewater produced from that industry. The basic principle is waste prevention by good housekeeping practices that will ultimately result in volume reduction and strength reduction. Industrial wastewater is treated the same way as domestic or municipal sewage—preliminary, primary, secondary and advanced treatment levels. Most of the treatment methods discussed is also applicable. There could however be peculiarities with different industrial depending on their major contaminant e.g. heavy metals, dye, etc. Industrial wastewater in Ghana is generated from breweries, distilleries, textile, chemical & pharmaceuticals and institutions and hotels which are mainly situated in Accra and Tema. In the Western and middle belt of Ghana, mining activities are predominant and the major polluter of our rivers. EPA-Ghana grants permits to industries and requires industries to install or build an in-house waste treatment plant. EPA-Ghana takes samples quarterly from these industrial wastewater plants for testing in their own laboratories for monitoring purposes. Most of those who have permit have treatment plants though not all of them are functioning properly.

In recent years, there has been a growth of small-scale industries in the fruits and food processing industries in the Tema, light industrial area which do not have the resources to build treatment plant. Most of these small-scale industries empty their wastewater into nearby drains without treatment. In Kumasi, the principal generators of industrial wastewater in Kumasi are the two breweries, a soft drink bottling plant and an Abattoir.

9.13 STATUS OF WASTEWATER TREATMENT PLANTS IN GHANA

The use of on-site treatment systems is quite extensive. Individual and community/residential based septic tanks are the most preferred. Septic tanks only partially treat sewage, and the effluent is still rich in organic material. The septic tank has to be emptied from time to time and the disposal of the septic sludge causes severe public health and environmental particularly in urban areas. Major wastewater treatment methods found in Ghana includes stabilisation ponds, trickling filters and activated sludge plants. According to a recent survey, there are 46 wastewater treatment plants in Ghana. More than half of all treatment plants in Ghana are in the Greater Accra region, mainly in the capital city of Accra and port city of Tema. Brong Ahafo and Upper West regions have no treatment plants at all. The stabilization pond method is the most extensively used with almost all faecal sludge and large-capacity sewage treatment plants using the method. Most trickling filters and activated sludge plants recorded have a low capacity and belong to private enterprises like larger hotels. Only about 10 of the treatment plants are operational and it is not clear if these plants meet the EPA effluent guidelines. This can be attributed to the fact that the conventional methods are energy dependent and also when the mechanical parts become faulty, the part has to be imported making it too expensive to maintain. Low-cost, lowtechnology methods are however manageable.

9.14 CHALLENGES OF WASTEWATER MANAGEMENT

Wastewater management though not technically difficult can sometimes be faced with socio-economic challenges. A few of the challenges are discussed below.

9.14.1 Infrastructure

Most often than not, wastewater infrastructure are not the priority of most politicians and therefore very little investment are made. It is however important to consider wastewater infrastructure as equally important as water treatment plant because almost all the water produced ends up as wastewater.

9.14.2 Pollution of water sources

Effects of wastewater effluent on receiving water quality is enormous, it changes the aquatic environment thus interrupts with the aquatic ecosystem. The food we eat contains carbonaceous matter, nutrients, trace elements and salts and are contained in urine and faeces (black water).

Medications (drugs), chemicals and in recent times hormones (contraceptives) are also discharged into the wastewater treatment plant. Discharge guidelines must be strictly adhered to. This will ensure sustainability of water sources for posterity.

The precautionary and the polluter-pays principles which prevent or reduce pollution to the wastewater have proven to be very efficient in the industrialized countries and should be adapted in developing countries as well.

9.14.3 Choice of appropriate technology

Because the economy of most developing countries is donor driven, funds for wastewater plants are mainly from donors. For this reason, they tend to propose the technology which should be adopted. For this reason, when the beneficiaries, take over the facility, its management of the operations and maintenance of parts become quite challenging as the technical expertise, power requirements etc are not sustainable.

9.14.4 Sludge production

Treatment of wastewater results in the production of sewage sludge. There must be a reliable disposal method. If it must be used in agriculture, then the risks involved must be taken into consideration. Due to the presence of heavy metals in wastewater, it is sometimes feared that agricultural use may lead to accumulation of heavy metals in soils thereby contaminating of yields.

9.14.5 Reuse

Effluents which meet discharge standards could be used for agricultural purposes such as aquaculture or for irrigation of farmlands. The challenge however is that if wastewater treatment plants are not managed and continuously monitored to ensure good effluent quality, reuse becomes risky.

BIBLIOGRAPHY

- Becerra TA, Wilhelm M, Olsen J, Cockburn M, Ritz B. Ambient air pollution and autism in Los Angeles county, California. Environ Health Perspect. 2013;121:380–6.
- Becker R, Henderson V. Effects of air quality regulations on polluting industries. J Polit Econ. 2000;108:379–421.
- Beelen R, Hoek G, van den Brandt PA, Goldbohm RA, Fischer P, Schouten LJ, et al. Long-term effects of traffic-related air pollution on mortality in a Dutch cohort (NLCS-AIR study) Environ Health Perspect. 2008;116:196–202.
- Behrendt H, Alessandrini F, Buters J, Krämer U, Koren H, Ring J. Environmental pollution and allergy: Historical aspects. Chem Immunol Allergy. 2014;100:268–77.
- Bentayeb M, Simoni M, Norback D, Baldacci S, Maio S, Viegi G, et al. Indoor air pollution and respiratory health in the elderly. J Environ Sci Health A Tox Hazard Subst Environ Eng. 2013;48:1783–9.
- Biggeri A, Bellini P, Terracini B. Meta-analysis of the Italian studies on short-term effects of air pollution – MISA 1996-2002. Epidemiol Prev. 2004;28(4-5 Suppl):4–100.
- Bourcier T, Viboud C, Cohen JC, Thomas F, Bury T, Cadiot L, et al. Effects of air pollution and climatic conditions on the frequency of ophthalmological emergency examinations. Br J Ophthalmol. 2003;87:809–11.
- Brook RD. Cardiovascular effects of air pollution. Clin Sci (Lond) 2008;115:175–87.
- Brucker N, Charão MF, Moro AM, Ferrari P, Bubols G, Sauer E, et al. Atherosclerotic process in taxi drivers occupationally exposed to air pollution and co-morbidities. Environ Res. 2014;131:31–8.
- Brunekreef B, Beelen R, Hoek G, Schouten L, Bausch-Goldbohm S, Fischer P, et al. Effects of long-term exposure to traffic-related air pollution on respiratory

and cardiovascular mortality in the Netherlands: The NLCS-AIR study. Res Rep Health Eff Inst. 2009;139:5–71.

- Cairncross EK, John J, Zunckel M. A novel air pollution index based on the relative risk of daily mortality associated with short-term exposure to common air pollutants. Atmos Environ. 2007;41:8442–54.
- Cakir, FY (September 26, 2005). Greenhouse gas production: A comparison between aerobic and anaerobic wastewater treatment technology. Water Research. 39 (17): 4197–4203. doi:10.1016/j.watres.2005.07.042. PMID 16188289.
- Calderón-Garcidueñas L, Mora-Tiscareño A, Ontiveros E, Gómez-Garza G, Barragán-Mejía G, Broadway J, et al. Air pollution, cognitive deficits and brain abnormalities: A pilot study with children and dogs. Brain Cogn. 2008;68:117–27.
- Calderón-Garcidueñas L, Solt AC, Henríquez-Roldán C, Torres-Jardón R, Nuse B, Herritt L, et al. Long-term air pollution exposure is associated with neuroinflammation, an altered innate immune response, disruption of the blood-brain barrier, ultrafine particulate deposition, and accumulation of amyloid beta-42 and alpha-synuclein in children and young adults. Toxicol Pathol. 2008;36:289–310.
- California Environmental Protection Agency. Operator Certification Requirements TABLE (PDF). State Water Resources Control Board. State of California. Retrieved March 30, 2015.
- Camargo JA, Alonso A. Ecological and toxicological effects of inorganic nitrogen pollution in aquatic ecosystems: A global assessment. Environ Int. 2006;32:831–49.
- Carugno M, Consonni D, Randi G, Catelan D, Grisotto L, Bertazzi PA, et al. Air pollution exposure, cause-specific deaths and hospitalizations in a highly polluted Italian region. Environ Res. 2016;147:415–24.
- Catcott EJ. Effects of air pollution on animals. Monogr Ser World Health Organ. 1961;46:221–31.
- Chang CJ, Yang HH, Chang CA, Tsai HY. Relationship between air pollution and outpatient visits for nonspecific conjunctivitis. Invest Ophthalmol Vis Sci. 2012;53:429–33.
- Chen B, Kan H. Air pollution and population health: A global challenge. Environ Health Prev Med. 2008;13:94–101.
- Chen TM, Gokhale J, Shofer S, Kuschner WG. Outdoor air pollution: Nitrogen dioxide, sulfur dioxide, and carbon monoxide health effects. Am J Med Sci. 2007;333:249–56.
- Chi CC. Growth with pollution: Unsustainable development in Taiwan and its consequences. Stud Comp Int Dev. 1994;29:23–47.

- Chung KF, Zhang J, Zhong N. Outdoor air pollution and respiratory health in Asia. Respirology. 2011;16:1023–6.
- Daniels, Jeff (2015-04-15). California drought gives 'toilet to tap' a new level of attention. CNBC. Retrieved 2020-04-27.
- de Cock M, Maas YG, van de Bor M. Does perinatal exposure to endocrine disruptors induce autism spectrum and attention deficit hyperactivity disorders? Review. Acta Paediatr. 2012;101:811–8.
- Dietert RR. Distinguishing environmental causes of immune dysfunction from pediatric triggers of disease. Open Pediatr Med J. 2009;3:38–44.
- Dincer I. Renewable energy and sustainable development: A crucial review. Renewable Sustain Energy Rev. 2000;4:157–75.
- Drakaki E, Dessinioti C, Antoniou CV. Air pollution and the skin. Front Environ Sci. 2014;2:11.
- Environmental Engineer. Science Buddies. Science Buddies, Inc. Retrieved March 21, 2015.
- Environmental Engineer: Career Profile. Study.com. Study.com, Mountain View, CA. Retrieved March 21, 2015.
- EUROPEAN SANITARY REFORM.; The British Sanitary Legislation, The New York Times, July 31, 1865
- Evans GW. The built environment and mental health. J Urban Health. 2003;80:536–55.
- Fairfax County Department of Human Resources. Wastewater Plant Operations Manager Classification Specification. Fairfax County Classifications. Fairfax County, VA. Retrieved March 21, 2015.
- Faiz A, Weaver CS, Walsh MP. Air Pollution from Motor Vehicles: Standards and Technologies for Controlling Emissions. 1st ed. Washington, DC: World Bank Publications; 1996.
- Fares S, Vargas R, Detto M, Goldstein AH, Karlik J, Paoletti E, et al. Tropospheric ozone reduces carbon assimilation in trees: Estimates from analysis of continuous flux measurements. Glob Chang Biol. 2013;19:2427–43.
- Ground Water Resource Mapping. New York State Department of Resource Conservation. State of New York. Retrieved March 23, 2015.
- Hanson, Harry G. (1956). Sanitary Engineering. Public Health Reports. 71 (2): 187–190. doi:10.2307/4589388. ISSN 0094-6214.
- Hardy, Anne (1984). Water and the Search for Public Health in London in the Eighteenth and Nineteenth Centuries. Medical History. 28 (3): 250–82. doi:10.1017/s0025727300035936. PMC 1139446. PMID 6390024.

- Heavy Downpours & Sewage Overflows. medialibrary.climatecentral.org. Retrieved 2020-04-27.
- Jha, Rajan (2016-08-30). Water pipes, infrastructure could buckle under climate change. GreenBiz. Retrieved 2020-04-27.
- Lattemann, Sabine; et al. (2008). Environmental impact and impact assessment of seawater desalination. Desalination. 220 (1): 1–15. doi:10.1016/j.desal.2007.03.009. Retrieved April 20, 2015.
- London's Victorian sewer system. Thames Water. Tames Water. Retrieved April 4, 2015.
- Matsuo, Tomonori; et al. Civil Engineering Vol. I (PDF). Wastewater Management Engineering. Encyclodedia of Life Support Systems. Retrieved April 3, 2015.
- Our Story. www.wm.com. Retrieved 2021-12-06.
- Rijsberman, Frank; Zwane, Alix Peterson (2012). WATER & SANITATION.
- Sanitary Reform of London, The Working Collection of Sir Joseph Bazalgette, British and Commonwealth Literary Studies, Green Library, Stanford University
- Schreiber, Melody (10 May 2018). The Housing Crisis in the Arctic, Caused by Melting Permafrost. Bloomberg.com. Retrieved 2020-04-27.
- Solid Waste Collection and Disposal. www.edcgov.us. Retrieved 2021-12-06.
- Steinbeck, SG. An Abridged History of Onsite Wastewater: Early Years to Present (PDF). REHS Onsite Water Protection Branch. Environmental Health Section, Division of Public Health, NC Department of Health and Human Services. Retrieved April 9, 2015.
- Summary Report for: 17-2081.01 - Water/Wastewater Engineers. O*net Online. U.S. Department of Labor. Retrieved March 23, 2015.
- Wages and Employment Trends. Onetonline.org. U.S. Department of Labor. Retrieved March 25, 2015.
- Waste Management San Joaquin Valley. www.wm.com. Retrieved 2021-12-06.
- Wastewater Treatment Plant Classification, Operator Certification, and Contract Operator Registration (PDF). State Water Resources Control Board. State of California. Retrieved April 20, 2015.

INDEX

T

V

W

www.ingramcontent.com/pod-product-compliance
Lightning Source LLC
LaVergne TN
LVHW062119200926
848484LV00007B/1718